Ann and Myron Sutton

A Chanticleer Press Edition

WILDLIFE OF THE FORESTS

HARRY N. ABRAMS, INC., PUBLISHERS, NEW YORK

Jacket. *A nocturnal long-eared owl (Asio otus) glides to a silent landing in a German coniferous forest.*

First frontispiece. *The wild boar* (Sus scrofa), *seen here in a German forest, is a familiar forest species in many lands.*

Second frontispiece. *This bush baby* (Galago demidovii) *of Central Africa, an endangered species, represents forest animals so delicately adjusted to their environment that prolonged disturbance threatens their survival.*

Third frontispiece. *A South American uakari* (Cacajao rubicundus) *leaps from a tropical forest tree. Such dramatic action, as well as endlessly varied sound and song, all fill the world's forests.*

Overleaf. *South American scarlet macaws* (Ara macao) *along a tributary of the Amazon River were photographed from the air. Such color—often iridescent and breathtaking—enlivens the shadows of woodlands or fills the sky above with an aurora of hues and patterns.*

Library of Congress Catalogue Card Number: 79–10874

Sutton, Ann
Wildlife of the forests.
(Habitat series)
A Chanticleer Press edition.
Includes index.
1. Forests and forestry. 2. Forest ecology.
3. Forest fauna. I. Sutton, Myron, joint author.
II. Title.
QH86.S9   574.5′264   79–10874
ISBN 0–8109–1759–9

Color reproductions by C. Angerer & Göschel, Vienna, Austria
Composition by Neil W. Kelley, Middleton, Massachusetts
Printed and bound by Amilcare Pizzi, S.p.A., Milan, Italy

Note: Illustrations are numbered according to the pages on which they appear.

Prepared and produced by Chanticleer Press, Inc., New York:
Publisher: Paul Steiner
Editor-in-Chief: Milton Rugoff
Managing Editor: Gudrun Buettner
Project Editor: Susan Costello
Assistant Editor: Mary Beth Brewer
Production: Helga Lose, Dean Gibson
Art Associate: Carol Nehring
Art Assistants: Johann Wechter, Dolores R. Santoliquido
Picture Librarian: Joan Lynch
Picture Assistant: Jill Farmer
Maps: Antonio Petruccelli
Design: Massimo Vignelli

Consultants:
General Consultant, John Farrand, Jr., The American Museum of Natural History; Appendix: Entomology, Raymond A. Mendez, The American Museum of Natural History; Plant Evolution, Richard Spellenberg, New Mexico State University

Contents

Preface

This book reaches into the heart of wild forests on every continent to present a firsthand look at the countless relationships between woodland animals and their environment. It covers the worlds of summer and winter, of day and night, of past and present, and of equatorial, temperate, and boreal regions. For there is not one forest but a thousand, each changing and vibrant with life, each containing the myriad marvels of a complex ecosystem.

In the pages that follow, we penetrate a tree to witness the hidden action there; then burrow into the soil for glimpses of an invertebrate world so immense and varied that man's analyses of it are still far from complete.

In the long-occupied lands of Europe we find, perhaps to our surprise, that natural forests full of wild animals still exist, and are filling again with something of the same ecological vitality that existed when Neanderthal man roamed through the woods and valleys.

Going eastward into Asia, where Marco Polo more than seven centuries ago wrote of wildlife conservation at the court of Kublai Khan, we come upon such seeming anomalies as tigers bounding through the snow. Because so much of natural Asia is already gone, a great deal of work is required to save or bring back its disappearing woodlands. Indicative of the growing sympathy for dwindling wildlife, sanctuaries are giving tigers and rhinos a respite from hunting. Thanks to such measures, we may today ride by elephant into great forest communities dominated by sal trees and happily witness the comeback of sloth bears, jungle fowl, boars, deer, and monkeys.

In Africa, where wildlife-based tourism has in places become the principal source of income, some Africans are making heroic efforts to ensure that huge national parklands will remain unspoiled for public use indefinitely. Once richly endowed with wildlife, Africa is now largely a domain of settled human beings and cattle. But in a number of equatorial rain forests, life goes on to some degree as it always has.

The Americas, from Alaska to Tierra del Fuego, remain in the vanguard of woodland preservation, and one may find extraordinary collections of living organisms in both North and South America. One of the most awesome forest phenomena is the deadly mass march of army ants in tropical South America. No other continent has such a striking array of colorful hummingbirds and butterflies, or an equatorial forest as vast as that of the Amazon Basin. To see the most primitive habitats of all, we take the reader by native dugout into the heart of this luxuriant, little-known natural world, almost as scientist-explorers did in their adventures more than a century ago.

Among the most distinctive of natural environments are the eucalypt forests of Australia, with the dazzling flocks of parrots, calls of lyrebirds, howls of dingos, and unique koalas and kangaroos. In few other regions has there been such marked progress in national park protection.

All this, then, is cause for excitement—and some hope—as we visit the endlessly changing natural forests of the world.

Ann and Myron Sutton

Introduction:
An Uproar of Life

Ever since the first adventurous trees, small though they may have been, began to evolve on land more than 400 million years ago, the forests of Planet Earth have been in constant change. Almost from the outset they provided shelter against wind, rain, sun, and heat. They provided food sources. They absorbed rainfall and stored fresh water. Forest fronds moderated the force of winds and rainfall and the burning rays of the sun. These useful qualities made them attractive habitats to other plants and to animals that developed on land, and in due course, the burgeoning forest systems gave at least some protection against hostile natural forces and enemies.

At an early period the forest must already have teemed with life. As new animals evolved, they made their homes in the soil or among the tree trunks, leaves, and branches. They nested and reproduced, each species changing and evolving with the passing millennia. We have had only a few centuries of scientific study to understand the orderly way these changes came about over millions of years.

Forests are among the most successful of all natural communities, as evidenced by their survival since the earliest eras of organic development. No other vegetational grouping is so widespread or diverse—or, unfortunately, is disappearing so rapidly.

Few forests today retain all their primeval characteristics, their original fauna and flora, or even the natural processes and interrelationships that made them what they were before human beings came on the scene. That is why, as we travel around the world to examine both the usual and the unusual differences among forests, we often visit a number of national parks and related nature reserves. For it is principally in such protected regions that we still find nature in a more or less *natural* state. These are, in fact, sanctuaries not just for trees but for all the life within them. In such controlled surroundings we can examine the current condition of continuously evolving plants and animals, continent by continent, and observe what changes are going on. While doing so, we probe the adaptive traits and secrets of the incredibly varied animals inhabiting the world's forests, in order to introduce the living ecology of natural forest communities.

This search for the survival process and pattern in nature takes us on a fascinating journey into forests throughout the world. And we soon realize these woodland communities are indeed, as the naturalist-explorer Henry Walter Bates once said of the Amazon forests, filled with "an uproar of life."

Early Forest Life
Botanists believe the first tree was probably less than a meter tall, with a tiny stem and no true leaves, but did have the beginnings of a woody structure that supported it in the wind and held it up toward the sunlight. When and where these first trees grew is a mystery. But by Devonian times, some 350 million years ago, huge forests had become established, not such as those known today but collections of fern-like trees called *Eospermatopteris*, or "dawn seed ferns." In the Catskill Mountains of New

12. *Like their ancestors in the world's first forests, hundreds of millions of years ago, giant ferns continue to thrive today, such as these in Victoria, Australia.*
15. *Coniferous forests, ranging from clusters of stunted polar spruces to immense pines and redwoods, have extended into a wide range of geographical areas and habitats. Shown here in the Sierra Nevada range of North America is a remnant patch of one of the woods that covered the continents millennia ago.*
16–17. *Deciduous forests, often mixed with coniferous trees, flourish in habitats where at least part of the year is warm and moist. In this New Hampshire scene, the passing of summer is signaled by changes in coloration and the falling of leaves, which mark the passage to the coming dormancy of winter.*

York State, the remains of a large late Devonian forest include fern trees up to 1 meter in diameter and 12 meters tall. Fossil plants of Lower to Upper Devonian age occur in widely scattered localities throughout the Northern Hemisphere. Early vascular plants such as *Germanophyton psygmophylloides* from the Lower Devonian of Germany and *Psygmophyllum gilkineti* from the Middle Devonian of Belgium suggest that plant life existed on Earth much earlier than is supposed. Tantalizing bits of evidence indicate that vascular plants were present as far back as Cambrian, and perhaps late Precambrian, times. In any case, Devonian deposits in Norway, the Spitsbergen Archipelago, South Africa, Great Britain, Ireland, Germany, Belgium, and elsewhere show that plant life had reached great complexity at an early era.

Another large Devonian genus was *Callixylon*, an early relative of the present conifers, which was known to have attained a diameter of nearly 2 meters. Members of the genus *Callixylon* were widely distributed, from the present location of the Arbuckle Mountains of Oklahoma to the Donets Basin of the Soviet Union. These trees were forerunners of the much more widely known Coal Age giants of Carboniferous times, the *Lepidodendron*, which were about a hundred times as tall (or about 30 meters) as their diminutive present-day relatives, the club mosses.

The ancient forests buzzed with a great number and diversity of insects, several with wingspans up to 70 centimeters. These hardly matched some of the efficient flying invertebrates of today, for they had rigid wings that could not be folded. The earliest insects had no wings at all, but they occupied their ecological niches apparently with great success: for example, they seem to have been well adapted to life in the forest, because half the known Paleozoic insects had piercing or sucking mouthparts for consuming plant juices. In turn, of course, these insects were food for the larger dragonflies.

Today insects represent over 70 percent of the million species of animals that have been described by scientists. Among the most abundant types (nearly 4,000 species) are springtails, primitive holdovers from ancient forests that now occupy the rich soil and humus of forest floors. In some places they are so abundant that as many as 2,000 of them inhabit each liter of soil—which means good eating for spiders, myriapods, and ground beetles.

Forest followed forest during the Paleozoic era, from about 425 to 280 million years ago, laying down so many layers of decaying vegetation that compression turned them into vast coal seams. Gradually the primitive plants of ancient forests—the seed bearers—diminished in competition with an emerging group of successful plants that had a new method of spreading their progeny. About 200 million years ago, early in the era of dinosaurs, seed-bearing cycads and conifers began to replace spore-bearing plants. By the time the first warm-blooded mammals appeared on earth, conifers were dominant and continued in that position for millions of years. Gradually the conifers gave way to another group of trees (broad-leaved), flowering and deciduous, which first appeared during the

reptilian age and which by 60 million years ago had spread to ecosystems around the world.

The trees and woods never ceased to shelter animals. Small insect-eating mammals of the Paleocene, insignificant in size, included primitive hedgehogs, shrew-like animals, and forerunners of the opossums. During the Eocene epoch, which lasted some 20 million years, several sloth-like creatures *(Metacheiromys)* evolved, and their South American descendants still survive as armadillos, anteaters, and sloths. Then squirrels, cats, birds, beavers, tapirs, rhinoceroses, and other forest dwellers succes-sively made their appearance, so that by the end of the Oligocene epoch, about 25 million years ago, animal life had begun to take on the forms we know today.

The Forests of Today

Modern forests have become established not by chance but owing to a variety of natural—and perhaps some unnatural—conditions. Global air circulation may influence where forests grow, simply because in certain latitudes dry air comes down from the upper levels and creates conditions under which only desert vegetation can survive. Topography, or configuration of the land surface, and the amount of sunlight that falls on certain landscapes may also decide the nature and distribution of forest. So do soils, their water content, and various other factors. If these conditions change, the forest also changes.

Forests cannot get up and flee in the face of ravaging storms or seek out water holes during drought; thus each tree must be adapted to extremes of its own environ-ment. Some trees, for example, have become almost "fire-proof," including certain pines whose bark may be scarred during a rapidly moving fire but otherwise suffer no damage. Some varieties of trees, such as conifers, are well adapted to cold, though many places on Earth are too cold even for conifers. All trees need water for suste-nance, but some leaves of trees that evolved in desert environments have become so adapted by means of reduced transpiration, or water loss, that certain types of forest can do very well in arid regions.

The result of such adaptation is a remarkable variety of forest ecosystems. In the Arctic, few trees besides stunted spruce and birch can grow in the earth-encircling band of forests called the taiga. But taiga animal life is not similarly stunted; the abundant moose and bear manage to find enough forage or prey to thrive.

The bands of forest change as we move into regions with warmer temperatures. Dense stands of spruce, fir, pine, and other large cone-bearing trees occupy the Northern and/or Southern Hemispheres to about 30° lati-tude. In these coniferous forests black bear, deer, and elk are dominant; fur-bearing mammals become abundant, and so do such birds as grouse.

As one proceeds southward from the taiga, the first important hint of a region of deciduous trees (trees whose leaves fall in autumn and grow anew the following spring) is the prevalence of aspen groves among the conifers. By the time we reach the subtropics, we have passed through a vast domain of deciduous trees: oaks, maples, alders, and

18. *Like a huge greenhouse, hot, humid, and sometimes hostile to man, tropical rain forests are rich in flora and fauna, typified by this lowland rain forest in Surinam. Hidden within the woods and waterways are myriad species of living organisms, a biological treasure of incalculable value to all life on earth.*

the like, alive with the music of songbirds and busy with the prowlings of boars, raccoons, and deer.

This sequence of forest belts from arctic to tropics can be interrupted by mountain ranges; but these montane forests themselves have various bands, with coniferous forests found at higher elevations and deciduous woods located at lower levels.

Air circulation also has great influence on climate, for the air either brings or withholds moisture. Perhaps the best example of this is the monsoon forest of Asia, where summer rains and mist sweep in and saturate the vegetation, from lowland sal forests all the way up to coniferous woods high in the Himalayan Range. Tigers, rhinos, monkeys, and even the high-altitude yaks seem to thrive on this cycle of all-wet or all-dry air.

Certain North American temperate forests near the sea may receive nearly 4,000 millimeters of precipitation spread evenly throughout the year. Deserts, of course, receive far less moisture; yet vegetation adapts and flourishes there despite the arid conditions. Giant cacti up to 20 meters tall may weigh as much as 9,000 kilograms and grow in densities of 150 per hectare. These spiny groves are alive with quail, peccaries, and reptiles such as rattlesnakes.

Hot, humid equatorial rain forests, with liana-covered trees crowded as closely together as possible, seem to typify tropical forest zones, but they are not so common and widespread as we may think. Africa and Asia have relatively little of this type of tropical landscape; the most concentrated such area in the Americas is in the immense Amazon forest. In Africa, tropical forest animals include vervet monkeys; in Asia, gibbons. Along with these are, of course, innumerable other arboreal animals. The woods shelter reticent antelopes or shy sloths.

And who knows how many millions of insects and other invertebrates, such as termites and army ants?

As the mountains rise from these rain-drenched lowlands, they support dense forests wreathed in clouds—forests filled with oaks and orchids, frogs and toads, and exquisite birds such as the Central American quetzal.

The rest of the tropics, however, is largely dry, though abundant with life. Across the vast miombo woodlands of Africa, among low *Brachystegia* and *Julbernadia* trees, roam sables. There are forests of thorn trees in southern Africa, groves of giant ferns in Australia, and swamp forests on nearly every continent.

Thus we find what seems like a different kind of forest ecosystem for every sweep of wind, every band of rainfall, every coastal lowland, every cordillera, every change in latitude. The remarkable diversity of plants and animals makes these zones endlessly fascinating. No human being has accumulated more than a fraction of the knowledge to be gained of these systems. And not for another hundred years, or a thousand perhaps, will humankind even be able to identify all the inhabitants of woods, much less learn thoroughly about their life cycles and their ecological relationships. Nonetheless, we can glimpse enough aspects of forest life to understand some of the laws of nature that account for the "uproar of life" within them.

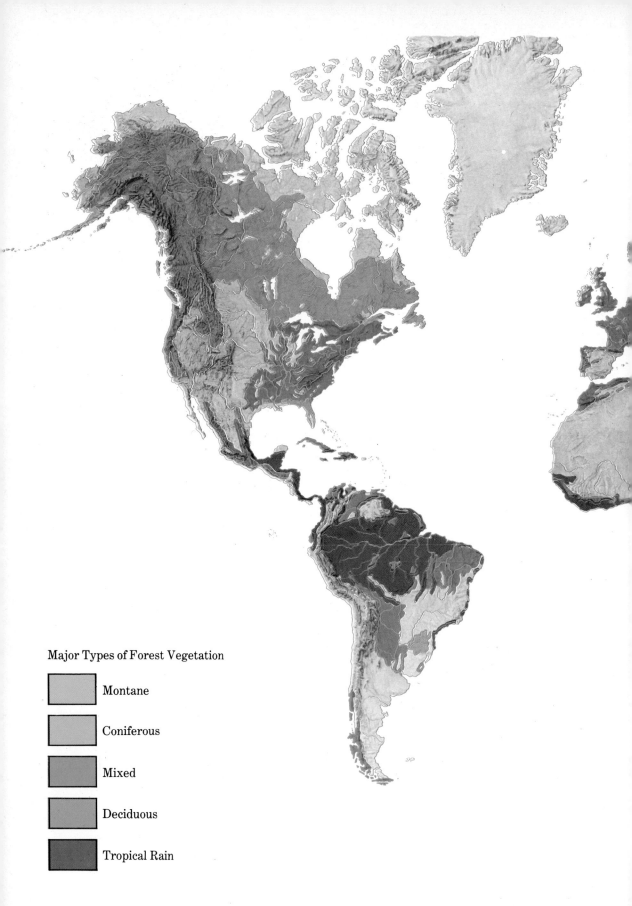

Major Types of Forest Vegetation

Montane

Coniferous

Mixed

Deciduous

Tropical Rain

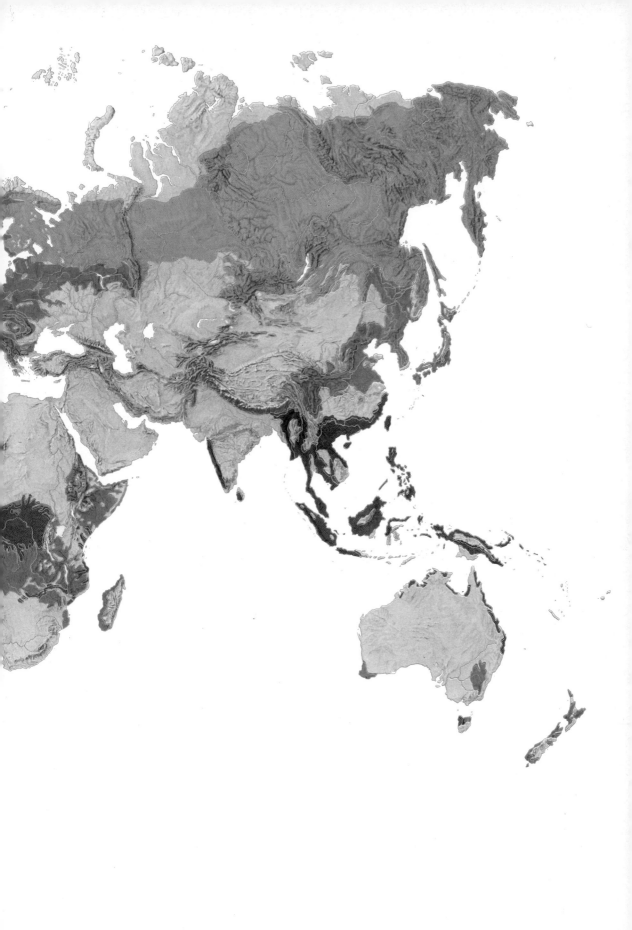

The Tree and the Forest

Contrary to appearances, a tree is an active, dynamic organism. Inside it—unseen except when we use a microscope—there is a two-way traffic of fluids that keeps each tree alive and supports an age-old growth and manufacturing process.

A tree's roots spread through the soil, deeply or widely, or both, depending on conditions above and below ground. If there is little water in the soil, roots may penetrate far down before they find enough moisture to sustain the plant's growth, or they may spread out close to the surface to catch rainwater before it soaks through. Trees that thrive where water is plentiful, as in river bottoms, tend to have shallow, wide-spreading root systems.

The root network is a tangle of lacy, living threads. A huge oak tree might have hundreds of kilometers of strands, looping, twisting, and reaching out in all directions. Root hairs thread among the grains of soil and thus help to anchor the tree against high winds.

The root cells also absorb water, and their structure is such that water travels from cell to cell. If the weather and soil are warm enough, growth proceeds rapidly: the cells divide, the root elongates and expands, the first leaves take shape, the tender stem begins to form, the tree enlarges as it rises higher and higher. Some tree-like flowerstalks such as the desert Agave, whose trunk is as thick in diameter as a human arm, grow upward at the rate of 1 meter every three days. Certain tropical vegetation where heavy downpours of rain are common does even better than that. But water is the common denominator of life, for trees as well as for animals.

Anatomy of a Tree

A tree's internal "plumbing" system may at first appear uncanny, enabling it to lift water for more than a hundred meters to the top of the tallest tree or out into widely spreading branches and thousands of leaves and then to carry liquids back again to the trunk and roots. There is, however, no magic involved; it is a marvelously efficient arrangement tried out and refined over millions of years of biological evolution.

Water containing dissolved minerals rises to the leaves, where sugar and protein are manufactured with the help of solar energy and then taken back down to nourish the stem and roots. This transfer happens principally in the trunk, and in most trees only near the outer surface of that trunk. A thin film of living cells called the cambium layer supports several kinds of action: the inside of this cambium sheath, toward the center of the tree, consists of a network of tubes, the xylem, through which water rises. Outside the cambium sheath, just under the bark, another system of tubes called the phloem carries food-filled liquids downward.

The cambium is a growing layer, producing sapwood inside and bark outside. With water ordinarily abundant in spring, the cambium produces numerous fibrous plant cells that harden into light-colored wood. During the rest of the year, with water usually less abundant, the fibers form more compactly to produce a condensed layer of darker-colored wood. All this growth occurs in cylindrical form, up

24. *Within its own enfolding bark, the tree can be like this massive valley oak* (Quercus lobata) *of California, an immense and productive engine of life. A tree such as this is a shelter or perch for birds, a home for insects, and a food producer for multiple forms of animal life.*

26–27. *This bromeliad perched on a tree branch in a tropical forest in northern Peru, by collecting and holding moisture, itself becomes a base of operations for tiny aquatic and terrestrial animals.*

28–29. *The woods are alive with organisms that in multiple ways transfer pollen from flower to flower, thereby helping to assure plant reproduction—and the continuation of their own food supply.*
Above. *For pollination, this tongue orchid* (Cryptostyles) *in Victoria, Australia, relies on an attempted mating by a male wasp* (Lissopimpla).
29 top. *A ruby-throated hummingbird* (Archilochus colubris), *with its elongated beak, feeds on a trumpet creeper* (Campsis radicans) *in Virginia.*
29 bottom. *A Peruvian bee* (Englossa) *has an elongated tongue adapted to reach nectar in deep flowers.*

and down the tree and more or less continuously around it. But when we examine a tree in cross section, these cylinders of growth appear as rings, normally light ones and dark ones, with each pair representing a year's growth.

Dendrologists studying these tree rings can literally look back into the past, to see at a glance whether the tree had large amounts of water in a certain year and vigorously produced sapwood, or had little and produced only condensed dark rings. That knowledge, in turn, gives us a fairly reliable idea of what the climate was like (whether the growing seasons were rainy, average, or dry), and tree scientists through such a system have devised a picture of world climates far back into periods before man kept weather records.

Obviously a tree must be constantly wet inside if it is to grow. Remove the water at the roots, and the tree dies. Girdle the tree by cutting into the bark all the way around, thus breaking the xylem and phloem tubes that transport life-giving water, minerals, and nutrients, and the tree dies. Some animals, such as porcupines, do girdle trees because they are fond of the sweet juices and flesh of the cambium sheath; in doing so, they kill trees, but seldom in such quantity as to harm whole forests. Beavers gnaw at and cut down whole trees, and they have wiped out entire groves of riverside woods. But the forest life cycle goes on, and woods grow back.

Prolonged drought may kill trees because there won't be enough water in the soil to supply the leaves. The xylem and phloem tubes dry up, so that photosynthesis, the making of carbohydrates in the leaf, stops. Leaves wither, brown, and fall, and the roots shrivel. The tree dies, rots, becomes a home for trunk-dwelling birds, mammals, and insects, and eventually disintegrates into a nutritious "dust" that enriches the soil.

When the rains return, new seedlings may grow from seeds that have been dormant for years. Those seeds themselves develop from flowers. Trees of different kinds produce different kinds of flowers: hanging clusters such as the catkins of willows or brightly colored displays of tropical *Tabebuia.*

To increase the range of distribution of their progeny, varied methods of seed dispersal have developed. Winged seeds such as those of maple and ash float on the wind. Flavorful fleshy seeds like those of wild cherries pass through the digestive tracts of birds or mammals, later to germinate where deposited. Or buoyant seeds such as coconut and mangrove may float long distances on ocean currents before coming to rest on a muddy shore and germinating. All in all, the tree is an extraordinary mechanism for survival.

Profile of a Forest

Whether living or dead, the forest is a source of food—and an arena for food-eaters. Warblers fly among the leaves in search of insects; among its flowers, insects such as bees search for nectar. Underground, burrowing amphibians and a host of invertebrates probe the decaying litter for food. Indeed, the forest community seems to divide

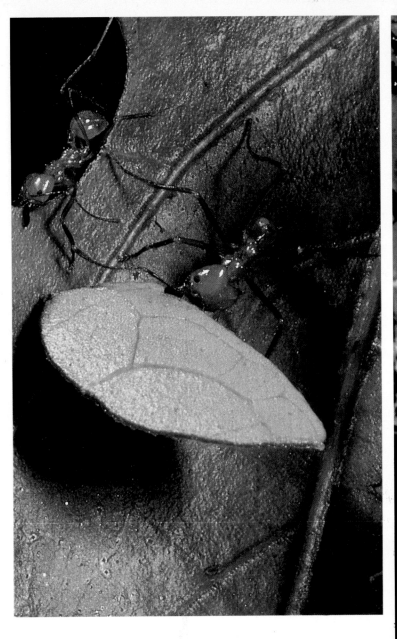

Leaf-cutter ants (Atta), farmers of the ant world, cultivate fungus gardens. Each ant species grows the particular fungus on which the colony depends. Worker ants cut out pieces of leaves (far left) and carry them back to the nest (left, top). Soldier leaf-cutter ants guard the entrance to the nest (above). Inside, the leaf fragments are chewed into a compost that nourishes the underground fungus.

naturally into levels of life, with the organisms found in each not readily interchangeable. Some birds and ants may circulate through all levels, from treetop to forest floor, but for the most part, animals prefer one ecological niche. We are no more likely to find earthworms in the treetops than honeybees among the roots.

This layering of animal life in vertical zones, though visible in nearly all woodlands, is best seen in rich deciduous or tropical woods. The lowest layer, composed of roots and soil, is a paradise for invertebrates and burrowing mammals. The bacteria and tiny fauna living in the earth are essential to the fertility of the forest floor. They stir and aerate it; they chew up dead vegetation and hasten the rotting process. Insects and millipedes eat the dead residue of plants that bacteria have already partially decomposed and eject it in smaller pieces. Thereupon, even smaller creatures break up the bits of humus and facilitate their further utilization by single-celled plants and animals. All this activity is sustained in a rather delicate equilibrium, and human beings who use chemical agents to rid gardens, meadows, fields, or forests of "destructive" insects may interfere with this vital process.

We become aware of such minute fauna when a firefly flashes at dusk or biting midges *(Heleidae)* assault our exposed flesh. But aside from such obvious indications, the soil abounds in springtails, amoebas, mites, centipedes, spiders, beetles, bugs, slugs, worms, and the larvae of other organisms. One of the most interesting of these creatures starts out as a scavenging grub that subsists for four years or more on rotting wood, then transforms into the dark brown, armored stag beetle *(Lucanus cervus)*. The adult lives for only a few weeks, feeding on sap, nectar, and fruit juices and seeking a mate. Because stag beetles reach 10 centimeters in length and look so ferocious, man has created something of a mystique around them. In France and Romania, for example, it was believed that when worn on a hat they could ward off evil.

Above the soil layer is the forest floor, where a different set of inhabitants thrives. Common ground dwellers are ungulate or hooved mammals such as deer. Although a deer is often an inhabitant of the forest edge, it leaps back into the interior of the woodland for shelter and protection. Its drab gray-and-brown coloration makes good camouflage, except when set against a backdrop of snow. The forest also provides emergency food; when nutritious grasses and forbs are unavailable, as when covered by snow, a deer may keep from starving by browsing on coniferous needles or perhaps tree bark (though the latter supplies more bulk than nutrition). In any case, the deer is a splendid example of the life found on the forest floor. In this same category one might include rodents, hares, boars, and ground-dwelling birds, which rely on grasses, seeds, and small herbaceous vegetation; plus such inevitable predators as snakes and foxes.

On this forest floor may grow a shrubby layer. Because many shrubs produce fruits, they are likely to be visited by such fruit-eating creatures as bears and birds. Shy animals also prefer the shelter of shrubby places, which offer some protection from preying hawks and owls.

Most minute forest organisms are seldom observed by man because they remain hidden in wood, soil, or the dark of night. But not so the fireflies of Southeast Asia: when males gather and light up simultaneously, the glowing spectacle may be seen for long distances through the forest, principally for female fireflies to see the brilliant display and be attracted to it.
Above. *A male Malaysian firefly* (Pteroptyx malaccae) *displaying.*
Opposite. *Countless male fireflies illuminating a* Sonneratia cascolaris *tree in Malaysia.*

The next-higher layer in our forest profile is composed of
small trees, such as some maples and hornbeams, that
survive in shade and thus do not have to compete for sunlight and open sky. Rain forests offer good examples
of lower layers overshadowed by a massive elevated
canopy of tree branches. Ferns, mosses, and lichens
grow thickly on trunks and limbs, along with epiphytes,
plants which are attached to trees but which get the
moisture and nutrients they need from the air and moisture
flowing past them.

The animals inhabiting this middle layer may be rather
versatile: they search the ground for prey and, since they
are climbers, may also hunt at somewhat higher levels.
Good examples of this community are certain cats such
as leopards and ocelots, which hunt from trees, leaping
down on their prey from the branches. Some cats may even
drag their victim back up into a tree to feast on it.

Animals adapted to climbing but comparatively heavy find
the sturdier branches of lower trees a more congenial
environment. Some bears and the larger arboreal iguanas
fit this category. But lighter-weight cats, such as ocelots
or margays, may climb higher to pursue such prey as
birds.

The layer of smaller trees merges into an upper layer of tall
trees, whose branches spread widely and intermingle as
they lift their leaves toward the sunlight. In a sense, these
taller trees compete with each other for open space. The
more sunlight they receive, the more they can manufacture
nutrients and gain strength. A few trees, called emergents,
are able to send shoots even above the highest canopy.
Fruits produced at these high levels satisfy the need of
certain animals for food and moisture, so that some
canopy-dwelling animals scarcely ever descend to the
ground.

Perhaps the best illustration of this layer of life is the
monkey community, particularly the nervous, vulnerable,
fast-moving species of the uppermost branches. These
slender upper branches also offer refuge from the big cats,
which may be too heavy and clumsy to climb so high.
Moreover, such a high abode enables the monkeys to escape
the many ground-dwelling predators, since, for example,
they never need to seek out a river to get a drink. But it also
means that they must find a forest where the branches
interlock tightly enough to afford continuous leaping and
climbing avenues to carry on their search for moisture-
providing fruits.

Evolution and adaptation to the upper story of the forest
has limited these fauna to existence on an almost horizontal
plane. There are still perils, of course, especially if an
animal carelessly exposes itself to avian predators from
above or accidentally falls and encounters a lower-level
predator such as the jaguar. But, just as the deer or boar
never climbs, many of these arboreal dwellers never
descend and are as suitably adapted to their special niche.
Other organisms also inhabiting high forest layers
include birds, reptiles, sloths, anteaters, gliders, and
such insects as moths and butterflies.

Though seldom mentioned in the literature, we can think of
one more level—the "sky layer"—clearly linked to

the life of the forest. This is the source of rain or fog and the home of soaring birds and flying insects, or even spiders that sail along on threads borne by the breeze. Steady or intermittent breezes ordinarily dry out the upper levels of a forest, or at least provide better overall circulation of air. Down on the less airy forest floor the greater humidity and the water dripping from upper leaves often support a community of plants and animals adapted to moist conditions. Exchange of pollen goes on in the open air, and winged seeds are carried by the winds to distant places, thus helping to expand and regenerate the whole living, interlocking forest ecosystem.

Eagles or hawks may be considered examples of animal life in the open layer above the forest. An airborne predator circles or soars on rising air currents and, at the strategic moment, folds its wings and dives in a silent, incredibly high-speed plunge after some rodent or reptile. Then it rises, prey held fast in its talons, to carry its meal off to a nest perhaps, or simply to the nearest snag or barren branch. There it presses the prey against a branch and tears off chunks of flesh with its powerful beak. Carrion-eating birds such as vultures may descend to pick at the remains of a dead fish or mammal.

Forest Openings and Edges

Many woodlands are characterized by openings, some of them more or less "accidental." Hurricanes and tornadoes, for example, may leave wide swaths when they destroy trees and wild animals in their violent passage. Other forest openings result from fields and clearings made by human settlers. If abandoned, however, such openings do not remain meadow-like very long. Flooded with sunlight, they nourish seedlings of herbs, grass, shrubs, and new trees, which all burgeon in a process called succession, with one plant or group of plants succeeding another as ecological conditions change— the more hardy crowding out the fragile and less adaptable. Eventually the trees grow tall, blocking the sunlight for all but a few smaller plants. In time the forest reaches a "climax," that is, a condition of stable, self-perpetuating growth. Usually what we see is the succession in progress rather than the climax, because trees sometimes take hundreds of years to grow to their maximum height.

The forest edge, or "ecotone," is important to many forms of life, and a daily "commuting" takes place between one area and another, as we have noted in the case of deer feeding in meadows and taking refuge in forests. In some cases the boundary between forest and meadow can be very distinct, owing to differences in soils or drainage or the keen competition among plants. Elsewhere, the ecotone can be a broad, gently modulated intermingling of meadows and trees, as where grasslands merge with forests. Whatever its extent, this intermediate zone is likely to contain plant and animal species of both vegetational types, along with others that have become adapted to life in such transitional areas. The "edge effect" results in a considerable density and variety of organisms, more so than in pure stands of forest or near the center of large meadows or savannas.

Layers of life: No ecological niche has remained unoccupied in the course of forest evolution. Animals with wings are among those which have adapted to the air and upper levels of trees, where they may subsist on fruit, nectar or leaves. One such example is the scarlet macaw (Ara macao) of Venezuela (34 top). Butterflies often concentrate lower down, on shrubs or the trunks of trees, as does Cairn's bird wings (Ornithoptera priamus euphorion), shown below, mating, in North Queensland, Australia. Reptiles, including the leaf-tailed gecko (Phyllurus platurus; 34 center), also inhabit the middle layer. On the forest floor one finds ground-dwelling birds, reptiles, and insects, together with such mammals as the garden dormouse (Eliomys; 34 bottom).

Contrasts

Some trees such as aspen, birch, pine, and mangrove grow in groves, clones, or geographic bands. Some forests have no distinguishable groves; instead, many species are thoroughly intermixed for thousands of square miles, as in Brazil. Some forests are sparse and arid; the North American ponderosa pine forest, for instance, is noted for its growth in open park-like meadows. Beech forests are open, too, since little other vegetation can get started in the wide tangle of roots and acid soil spreading out from the base of this tree.

One characteristic of temperate forests is that only a few of the species present are common and conspicuous in any one locale; other species within that community are found only in limited numbers. Thus we speak of "oak forests" or "pinewoods." Such prevailing species amount to "ecological dominants"—a situation that seldom occurs in tropical rain forests.

Some forests may be extremely dense and varied. When we asked Brazilian authorities how many species occurred in their luxuriant forests, they responded with a hopeless shrug. The Amazon forest is so immense, with seemingly infinite species of trees, and is as yet so little studied that no one knows the answer to such a question.

Some species are rare, such as cedars of Lebanon or *Araucaria* and certain quebrachos of southern South America. Some trees such as the franklinia *(Franklinia)* of North America are extinct in the wild state. Other trees never grow very large, as in the high-altitude rain forests of El Yunque in Puerto Rico. The largest living organism is the General Sherman tree, a giant sequoia *(Sequoia-dendron giganteum)* in California: 83 meters tall, 11 meters in base diameter, 31 meters in circumference, 635 metric tons in trunk weight, and 1,416 cubic meters in volume. This giant is estimated to be between 3,000 and 4,000 years old.

As for other forest contrasts, we know that certain trees thrive in fresh water, whereas others subsist in salt water; some forest systems have sparse life, while others have high population densities.

The Forest Ecosystem

Standing amid a forest somewhat resembles standing on a busy street. The forest does not have the same noise and pollution, but the comings and goings of its inhabitants all relate in some way to the total system. For example, standing in a temperate deciduous forest, we might observe a squirrel scampering down a tree trunk with a nut in its mouth, en route to burying that seed in some distant place—where it will later sprout and grow unless the squirrel reclaims it.

In a sense the forest is like a multi-dimensional chess game, with each layer having its own characters related to one another, and every layer, in turn, related in diverse ways to the levels above and below. If you move just one piece in a chess game, the whole play may be affected. Similarly, if but one part of the forest's interconnected ecosystems is altered, the evolutionary progress of the whole community may never be the same again.

Major Types of Forest Vegetation

- Montane
- Coniferous
- Mixed
- Deciduous
- Mediterranean
- Taiga

The Woods of Europe

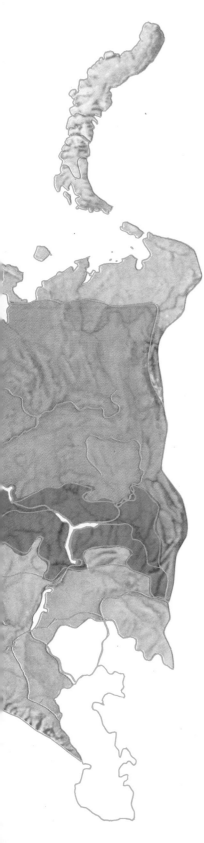

Majestic red deer in forest clearings? Swift owls diving out of trees? Great herds of wisents in the woods? These and other less dramatic creatures have survived over the centuries and still exist in Europe's diverse wooded regions.

As the map of Europe indicates, there are distinctive bands of woodlands stretching across the continent. South of the arctic barrens and the sparse vegetation of the tundra are the tenacious conifers of the taiga, clinging to life in Russia and Scandinavia. Well-adapted to ice, snow, and wind, these hardy survivors are dominated by species of great spruces and firs. A little farther south, where the habitat is less harsh, the conifers become denser, forming a thick mantle of boreal forest extending across northern Europe. Then, in warmer climates, conifers become far less conspicuous or are absent altogether, and there are forests of broad-leaved (deciduous) trees. In southernmost Europe, on the shores of the Mediterranean Sea, the land becomes almost desert-like, and the trees found there have evolved a capacity to endure weeks or months with scant rainfall. Broadly speaking, the continent's forests are banded horizontally across the landscape as if in general conformity with latitude.

Survivors of the Ice Age

That the forests themselves have survived the perils of European prehistory and antiquity is remarkable, for environmental hazards existed long before human beings appeared there in any great numbers. The woods of Europe underwent extensive modification for millennia, all from natural forces. None of the three principal vegetational bands—the taiga across the north, the central temperate deciduous region, or the Mediterranean scrub—escaped such natural processes completely, and the northernmost two zones were devastated.

The forests of Europe have, in fact, had a much more difficult time than those of other continents. One reason is that landforms hindered plant survival when continental glaciers advanced. Whereas in North America the mountain ranges formed in a predominantly north-south direction, thereby allowing plant and animal species to retreat southward during glacial advances, the mountains of Europe, from the Pyrenees to the Caucasus, have an east-west orientation that blocks any such southward retreat. That is why many European forests perished during the Ice Age. Worse yet, if any species did successfully cross those mountains and move toward the tropics, it was confronted with the formidable obstacle of the Mediterranean Sea.

Until glacial times, the temperate vegetation of North America, Europe, and Asia had been strikingly similar (as some of it still is). But after the ice melted, Europe had only some 35 genera of trees left, whereas China's vegetation remained nearly intact. One saving factor, in some ways inexplicable, is that certain places such as the Pieniny Mountains of Poland and Czechoslovakia and the driftless area of the central United States were surrounded by but never covered with ice. Nor were the Great Smoky Mountains of the southeastern United States, which are

a bit more southerly; there the temperate trees found a refuge from the ice and were able to continue their millennia-long evolution. Europe was not so fortunate.

The Owls: Birds of the Night

Since ancient times owls have been symbols of wisdom—perhaps because of their solemn appearance—but have also been considered birds of mystery and misfortune because of their nocturnal habits, eerie calls (from screeches to hoots and moans), and noiseless flight.

Of the thirteen species in Europe, some are found only in the great conifer forests of the taiga, but many occur in mixed woods such as the Black Forest. Owls are birds of prey but differ from other raptors in that they hunt by night. They are such effective hunters because of several physical features: their eyes, like those of many nocturnal animals, are huge; but even more important in locating prey at night is their hearing, so keen that some species seem able to attack even in pitch darkness. Both hearing and sight are aided by the peculiar, disk-like shape of an owl's face, which may well serve to concentrate incoming sensory impressions.

Most owls are also equipped with very sturdy feathered legs, powerful claws, and a strong hooked bill. When threatening or attacking another predator, the owl spreads its wings and expands its feathers until it appears to be twice its normal size. The ears, ordinarily only tufts of feathers, are raised and add to the "threat display."

Owls lay from one to a dozen eggs, depending upon the species, with the birds of warmer climates laying fewer eggs and arctic species producing the largest clutch. In what is probably a survival mechanism, owls lay their eggs at intervals and begin incubating as soon as the first egg is laid, so that the nestlings are of different ages. This "staggered birth" reduces the need to feed a large group of nestlings all at once, but it also means that in a time of scarcity the last-born may starve.

Even more remarkable, and perhaps another survival mechanism, is the way certain owls such as the hawk owl *(Surnia ulula)* and the snowy owl *(Nyctea scandiaca)* lay a large clutch in a year of abundance and a small clutch or no eggs at all in a poor year.

Of the owls found in Europe, the pygmy owl *(Glaucidium passerinum)* is the smallest, and the eagle owl *(Bubo bubo)* the largest. The pygmy, though smaller than a starling, hunts other birds and is active both day and night. It is recognized by its small size and the habit of flicking its tail. The eagle owl, by contrast, is a huge, almost eagle-size bird, which frequents dense forest and rocky gorges. Though a powerful hunter, capable of bringing down prey as large as a capercaillie and, according to one report, a 30-pound roe deer, its diet generally is small mammals. Its numbers have been greatly reduced by excessive hunting and the taking of eggs by collectors.

The Climate

Like the owl, other animals and plants must also endure the sometimes seemingly erratic patterns of European climate. Burrowing and hibernating animals manage to avoid extremes by living beneath ice or soil where the temperature seldom gets much below freezing. Other animals and plants are vulnerable to the elements. Heavy storms, record snowfall and rainfall, and devastating hail, not to mention prolonged drought, must be endured by forest organisms. Winds can be treacherous: the foehn is a gale that results when air masses ascend southerly mountain slopes, such as in the Alps, thereby cooling, dropping their moisture as rain or snow, then pushing up over the summit, descending the northerly slopes, warming again by compression, and causing odd temperature inversions. This wind, heated independently of the sun, becomes so dry that it desiccates everything in its path. If this persists long enough, wood can be ignited with the simplest spark. For all its parching capacities, the foehn can be beneficial in driving out winter cold, melting snow, and touching the European countryside temporarily with spring. But its peculiar action adds to the unpredictable environmental conditions to which plants and animals must adapt or perish.

As for other inconsistencies of European climate, one should realize that going north does not always mean arriving at a colder climate. Despite a substantial difference in latitude, Bergen, Norway, and Lyon, France, have nearly the same average January temperature. The sunny Mediterranean can get so cold that water freezes in canteens left outdoors. All we may be fairly sure of is that the farther we get from the moderating influence of ocean air masses, the colder it gets. That explains why northeast Europe—from Moscow north—is coldest.

In Europe topography favors east-west air movements, but north-south air movements are much more impeded by the almost unbroken line of mountains from the Pyrenees to the Caucasus, which blocks, slows, or diverts northward movement of tropical air and southward movement of polar air. That is one reason we encounter such balmy climates south of the great mountain ranges and usually such temperate or cool climates north of them. This partly explains such winds as the foehn.

To complicate matters, large high- and low-pressure systems—all outside Europe—determine the weather pattern from month to month and from place to place. The wettest localities are on the uplands and along a number of seacoasts; the driest are in the Arctic, southern Russia, and certain regions of Spain. In a general way, we can simplify the picture of the European weather pattern by saying that western Europe, including the British Isles, is mainly a maritime climate; south-central Europe is transitional between maritime and continental; Soviet Europe and northern Scandinavia are continental; and the Mediterranean region is subtropical.

Forests clearly reflect these weather patterns, although a glance at a map is enough to remind us that human influences did more than practically all natural factors combined to modify the location of forests. Of the great mixed deciduous-coniferous forest that once graced the midband of Europe from Britain to the Ural Mountains of the USSR, only unconnected patches remain. Most Mediterranean woodlands were long ago cut down. The only sizable natural European forests still existing are found in regions inhospitable to man: the boreal forests of far northern Scandinavia and the USSR. In colder northern climates hardy spruces thrive, along with their associated fur-bearing mammals, including bear and reindeer. Where relatively mild maritime air moves inland to the east, we find broad-leaved trees dominant, though mixed with pine and spruce, and only at higher altitudes do we encounter extensive coniferous forests. The broad-leaved trees cannot survive the extremes of cold in the far north or the dryness of the far south. Thus beech, elm, maple, and oak characterize the woods of the central portion of western Europe. Deer and fox roam these areas in rather wide ranges because they have more or less successfully adapted to human encroachment. Badgers and wild boars have similarly accommodated.

Around the Mediterranean the climate is generally dry—salubrious for man, but not necessarily for plants. The plants require a host of adaptations that allow them to grow where broad-leaved trees of more northern climes cannot. Shrubby thorn thickets, a habitat called maquis, occur in patches; their leaves are small, leathery, sometimes spiny, and often coated with a waxy layer. These adaptations help reduce water loss from plant surfaces. In spring, when rain does occur, the maquis plants bloom exuberantly, only to dry up in summer, their reproductive work done. If you have traveled in the wilder places around the Mediterranean, you surely will have noticed the aromatic flora.

Maritime Forests

Going east from the British Isles, we find ourselves in a wide irregular band of temperate broad-leaved trees extending through France, Denmark, Norway, and southern Sweden to Russia. For the most part, this is a gentle environment, subject to the warming influence of the sea. East of the maritime environments, the continental weather becomes too harsh for the survival of such broad-leaved trees. Thus, paradoxically, Europe's major forest has a delicacy and fragility that means it can suffer badly from misuse by human beings—yet a durability and an adaptability that helps it in withstanding natural ravages.

The trees most commonly resident in this zone are English, or pedunculate, oak *(Quercus robur)* and sessile oak *(Q. petraea)*, especially where good soils prevail. On more calcareous soils beeches grow, together with common ash *(Fraxinus excelsior)*; these two types, which also tolerate cool temperatures, may be found farther north or higher up on mountain slopes. On still higher slopes or in more northerly latitudes live the durable birches. Within these general environments we can look for less abundant trees in quite specific locations: alders, poplars, and willows near water; maples on better-drained soils; and elms in fertile places.

As the maritime influence diminishes farther east and north, the climate becomes more rigorous. In the USSR the coniferous species mix with deciduous woods. The taiga extends to roughly the latitude of Leningrad, and a zone of mixed forest stretches southward. Around Moscow the

Forest fires such as this one in a Portuguese pine woods (top) or another near an English forest (above), as well as high winds, rock slides, avalanches, and other natural phenomena, all alter forests throughout the world. With man's unceasing cutting and burning, the world's forests currently suffer a substantial net loss each year.

weather can be severe; snow covers the ground from late November through mid-April, and even in July temperatures can fall nearly to freezing. But farther south, environments moderate and the climate becomes rather uniform.

Cloudy and rainy weather suits the types of trees that have evolved in these temperate forests, principally mixtures of spruce and oak. In certain areas, the common hazel (*Corylus avellana*) and hornbeam (*Carpinus betulus*) grow along with them; but stated simply, this region of intermingling is the southern limit of spruce and the northern limit of oak. Though these woodlands may be dominated by oak, walking through them one will recognize such other familiar species as aspen, pine, linden, elm, and maple. If we explore the forests quietly and seek out more remote locales, we have a fair chance of observing elk, bear, wolf, and roe deer. Beyond this larger wildlife, there is a considerable fauna of smaller animals: fox, lynx, and ermine, for example. Typical woodland birds of the temperate forests include the pied flycatcher (*Muscicapa hypoleuca*), and wryneck (*Jynx torquilla*).

The oak-dominated European and Russian mixed forest extends no farther east than the Ural Mountains; yet after a gap of nearly 7,000 kilometers, oaks and filberts (though of different species) appear again in the Amur River basin of Siberia, on the Chinese border.

A Gentle Land: British Woods

By contrast—and Europe has many of them—the woods of Britain have a great supply of moisture, and hence a greener aspect than the arid Mediterranean scrub. They are much more northerly, however, and nature places limitations on them different from those placed on desert flora.

Perhaps the most severe of these intrusions was periodic devastation by advancing ice sheets; yet forests recovered with marked vigor. As the last glacial ice melted away from Britain and Europe some 11,000 years ago, tree species migrated back and established themselves rapidly. But about 5,000 years later, the British Isles became separated from the European mainland, and only those trees that had made the crossing from the Continent survived. Welsh native woodlands once consisted of alder, birch, and pine, which about 3000 B.C. grew in many places along the coast. But then a cooler and wetter climate set in, ending the coastal forests and relegating the alder and its associated species to peat deposits. Adverse climatic changes wiped out the higher British forests during prehistoric times; and later, about a thousand years ago, the once richly wooded lowlands in valleys and along streams were cleared for farms.

The remaining temperate deciduous woods of Europe are populated by scores of species of birds, many of which nest in the forest canopy. However, British birds depend more on shrubs than on trees, perhaps because the first birds to arrive after the retreat of the glaciers—not so long ago geologically—occupied the smaller, poorer woods and have not yet grown accustomed to the taller, canopy-making trees. Then, too, the early Saxons and Normans gradually

The world's 290,000 species of beetles, the Coleoptera, present a remarkable variety of forms and structures. Of their two pairs of wings, the anterior pair forms horny coverings no longer used for flight. Beetles inhabit nearly all types and parts of trees: bark beetles (Scolytidae) tunnel into a tree's cambium layer; larval forms of stag beetles (Lucanidae) live in decaying logs. 46 top. During their few weeks of adult life, greater stag beetles (Lucanus cervus) often fight with their huge mandibles, as is shown here. Bottom. A pair of rhinoceros beetles (Oryctes nasicornis) are observed in a German woods. Both male and female possess a horn, but that of the male, at right, is larger.

destroyed the original woodlands that were trying to
become established on the ice-scoured terrain. By Tudor
times, few woods were left except protected royal forests
and those on the estates of noblemen.

When extensive reforestation was carried out from the
17th to 20th centuries, thousands of hectares were planted
largely with foreign trees: spruce, larch, and Douglas fir.
The result of this is likely to be an alteration of bird
populations; already the crossbill *(Loxia curvirostra)*,
which eats only the seeds inside cones, nests in increasing
numbers in the conifer stands of southern England.

Other changes in bird populations will eventually follow
as planted woodlands grow up and replace the native ones.
At present, broad-leaved trees predominate in the
southern half of England, while conifers cover parts of
Scotland, Wales, and northern England. However, the
only coniferous tree that forms natural woods in England is
the common yew *(Taxus baccata)*; Wales has no natural
coniferous woods. European beech came rather late to
England, about 2000 B.C., but has become the typical tree
of the chalk escarpments to the south.

British woods of oak, gentle but busy, shelter robins
(Erithacus rubecula), chaffinches *(Fringilla coelebs)*,
willow warblers *(Phylloscopus trochilus)*, starlings
(Sturnus vulgaris), wrens *(Troglodytes troglodytes)*, and
blackbirds *(Turdus merula)*. Very often the forests of
Britain drip with fog and drizzle that wet the ferns
and mosses, form droplets on lichens attached to tree
trunks, and weaken a trifle more the rotting logs or stumps.
Yet rains will scarcely dampen the spirit of the mistle
thrush *(Turdus viscivorus)*, as it sings with gusto among
the dripping trees.

And nothing but the severest weather—and perhaps not
even that—will prevent the woodland birds from diligently
gathering twigs or fibers to weave their nests. Some
birds build no nests, of course, and others occupy holes in
tree trunks, sheltered at least from tempests. But the
remarkable, delicate structures of such species as the long-
tailed tit *(Aegithalos caudatus)* may be built by males
and females working together for two to three weeks,
utilizing a variety of materials.

The sporadic showers sometimes seem to go on for days,
but if this vexes human visitors, it supplies abundant
streamlets and enriches growth. To know English forests
intimately is to be enveloped by a montage of impressions:
cushions of moss, tangled roots, singing birds, circling
bats, darting woodpeckers, and sweeping clouds
of mist. The contrast of softness, haziness, and moisture
here with the hard gray rocks and dry shrubs of the high
Mediterranean cliffs is a picturesque part of the natural
scenic charm of Europe.

Another feature is the majestic aspect of woodland life
contributed by one of Britain's—and Europe's—most
stable and dramatic large mammals, the red deer.

Bold Bugler: The Red Deer
One of the largest and, as you will find if you get very close
to it, most dangerous animals of the European uplands is
the red deer *(Cervus elaphus)*, whose massive antlers can

As in forests throughout the world, these birds of British woodlands show a decided preference for certain foods and nesting sites.

48. *The crossbill* (Loxia curvirostra scotica), *favoring coniferous forests, has crossed mandibles that enable it to crack open pine cones for the seeds.*

49 top. *The mistle thrush* (Turdus viscivorus) *generally places its nest higher in a tree than do the other English thrushes. The female incubates its three to five eggs for about two weeks; then both parents feed the young.*

Center. *The most abundant summer visitor among warblers in Northern Europe is the willow warbler* (Phylloscopus trochilus), *common in woodlands and shrubby areas. These insectivorous warblers devour and feed to their young the grubs, worms, and other soft-bodied food available here only in summer.*

Bottom. *A fairly common yet rarely observed bird of Northern Europe is the woodcock* (Scolopax rusticola). *Its camouflage, resembling the coloring of dead leaves, helps it blend with the forest floor, where most of its life is spent.*

50–51. *Cruising low above hedges or through woodlands, then pouncing on small birds and mammals, the sparrow hawk* (Accipiter nisus) *is an important European predator. While the female, shown here, is incubating eggs, she and her brood are fed by the male.*

be wielded with the same agility as those of the North American wapiti it resembles. In central Spain, as the deer were uttering their mating calls, we walked among them at dusk (probably a bit too unmindful of the perils involved) on a November evening when the poplar trees blazed full yellow. Compared with the bugling of the wapiti at rutting time, the sound of the red deer seemed hoarser.

Our knowledge of the red deer depends heavily on studies made in Scotland by the scientist-conservationist Frank Fraser Darling. Rutting is a stimulating time, and with the coming of September the buck literally steps up his pace to a constant trot. His neck swells, and he wallows in the dust. He eats little but moss. About November his roars and bellows taper off to something of a bleat. He gathers the hinds, fends off newcomers, mates, then goes off to the mountains to recuperate. Yet for bucks and hinds alike, life can be distressing. These animals are often so bothered by flies that for relief they charge up out of the lowlands to the cooler mountains. Normally their preference in food is grass, ashes when they can find a burnt-over area, and even cast-off antlers because of their need for calcium.

Like most large wild animals, red deer seem to need considerable open space to live contentedly, even where food is plentiful. It has been calculated that each deer needs 16 hectares to itself; this is an average figure, for these animals spend most of the year in grazing herds of males and females. Even the herds mix: Darling found that red deer have developed a degree of sociability that is extraordinary among animals and that the continuation of the species depends on this.

Spruce and Silver Fir: The Black Forest
In southwestern Germany, where it curves in a broad arc from west to north, the Rhine borders the lower reaches of a hilly, even mountainous terrain (up to 1,508 meters) of dark coniferous woods called the Black Forest. This collection of coniferous groves, deciduous patches, attractive lakes, rocky crags, waterfalls, and settled areas occupies a broad stretch of Germany, from Stuttgart on the north to Lake Constance on the south—a distance of about 120 kilometers. Poets have compared its rolling contour to the heaving surface of a troubled sea. From among these heights the Danube River begins its 2,840-kilometer journey to the Black Sea. On a clear day, to the south, one can see the peaks of the Swiss Alps.

In the woods' natural areas may be seen the wildcat *(Felis sylvestris)*, which hides by day and feeds on small mammals and birds at night. By contrast, pine martens *(Martes* spp.), members of the weasel family, hunt during daylight, subsisting on a varied diet from squirrels to berries. Huge black woodpeckers *(Dryocopus martius)* hack open rotting logs in search of insects.

The forest recalls its wilder aspects when the male black grouse *(Lyrurus tetrix)* displays from mid-March to mid-June on a circular display ground inherited from its ancestors. This ceremony, a challenge to other males, usually takes place in early morning or late afternoon. Because of its limited occurrence, it is seldom witnessed by

*Only during the rutting season does the male red deer (*Cervus elaphus*) raise its voice—a roar or bellow that can carry for long distances. During the rut, male red deer become extremely active, as they battle with other males to possess more females and herd their harems, wallow in mud, and eventually mate.*

Above. *Red deer stags spar at the edge of a German forest. The one at right is caked with mud from wallowing.* Bottom. *A red deer stag, continuously vulnerable to challenge by other males, chases its hind.*

visitors to the forest. Much more ubiquitous, and more often seen, is the familiar roe deer.

All these species, in addition to foxes, capercaillie, owls, and other wild animals, suggest that the Black Forest has retained some of its natural ecosystems.

Master of Survival: The Roe Deer

Because so much of the original deciduous forest of Europe has been cleared and the rest modified by human activity, a report by a United Nations agency states that "the original species composition and ecological associations can hardly be ascertained."

Beech has in places become the dominant species, forming almost pure stands in Germany and Czechoslovakia. Over widespread areas, however, the forests of European lowlands and low hills contain mixtures of English oak and various species of birch, with a heavy intermingling of conifers, mostly Scotch pine *(Pinus sylvestris)*, spruce *(Picea excelsa)*, European larch *(Larix decidua)*, and fir *(Abies excelsa)*.

One of the most common animals of Europe's mixed forest-meadow systems is the little roe deer *(Capreolus capreolus)*, weighing up to 30 kilograms. Originally it was an animal of the forest, but with so many trees cut away and lands cleared, it had to adapt to habitats at the forest edge and in open fields. Uncannily, the animal has come to time its feedings so they occur when human beings are generally absent from the fields and woods: during the early morning and even at lunchtime. The other animals with which roe deer are associated—red deer, bears, wolves, and lynxes—did not adjust so successfully, and their populations have diminished. Their sizable decrease left the field open to the roe deer, which increased so greatly in numbers that nowadays hunters in the Federal Republic of Germany alone slay 500,000 of them each year.

This abundance seems a bit strange when one considers that the roe deer is not as free in its mating habits as the red deer. The buck usually mates with only one doe each year, and will fight vigorously to keep her. Furthermore, when the female is fertilized in July or August, the egg lies dormant in the uterus for more than four months before the young begins to develop. This system obviously succeeds in maintaining a large population of roe deer, which occur in nearly all of Europe and Asia and in most habitats, including fields and croplands.

Home in the Trees: A Red Squirrel's Life

Another masterpiece of ecological adaptation is the red squirrel *(Sciurus vulgaris)*. Red squirrels inhabit wild and managed forests from western and southern Europe to Scandinavia and Siberia, and the genus *Sciurus* consists of 190 additional species and subspecies in other parts of the world. The squirrel ranks high among the animals most successfully and harmoniously adapted to life among trees. Though human visitors to the forest hear its chatter and see it climbing tree trunks with ease or leaping fearlessly from limb to limb, most scarcely realize the full extent of its dependence on trees—and vice versa.

56 top. *The wide-ranging roe deer* (Capreolus capreolus), *seen here in a German forest, is the only deer with delayed implantation: the mother's fertilized egg lies dormant in the uterus for about four months.* Bottom. *When it is time for birth, the doe chases away her young of the previous season and retires to a forest thicket, where she may produce twins or, more often, a single new fawn like this day-old offspring.*

Virtually everything it eats comes from the immediate environment: sprouts, buds, blossoms, fruit, nuts, sap, and seeds, especially the seeds of fir and yew. We can readily tell where all this happens, for the squirrel is anything but tidy and haphazardly discards the debris from its collecting expeditions. Wherever a pile of scattered hulls and chaff covers the forest floor around the base of a tree we can be reasonably certain that the tree has produced an especially tempting food crop, much of which the squirrel has eaten or hauled away.

But its venturesome trait leads the squirrel to sample other local fare as well, including mushrooms, snails, insects, ant pupae, and perhaps even the eggs and young of nesting birds. Thus the squirrel's food cycle lies wholly within the forest ecosystem, as does its reproductive cycle. Squirrels construct a circular roofed nest of twigs, leaves, and moss, as much as 50 centimeters in diameter and up to 30 centimeters high. In bad weather this refuge may shelter the occupant for days at a time. An industrious squirrel builds a second home for sleeping or hiding purposes, or may cleverly fix up an abandoned bird's nest and move in.

If these entrepreneurial activities seem intelligent, it is not only because the squirrel has had to sharpen its survival instincts against martens, hawks, and other enemies but also because the essential raw materials are there in abundance. About the only commodity supplied from outside the forest ecosystem—besides solar energy, of course—is the water obtained from streams. Red squirrels live near water, for they must drink it; if their water source freezes in winter, they may eat snow. Food supply in winter is not a problem, providing the squirrels can remember where they so prudently stored the seeds of summer.

This very act of storage, and of forgetting, in its own way constitutes a return gift to the forest. Squirrels clearly, if unwittingly, help to plant new trees and thus spread the forest—a natural benefit to trees as well as to other forest animals, erosion control, and so on. Such a circle of life, in which the squirrel gets and gives, is only part of countless other life cycles that compose the complex woodland ecosystem.

Among the most familiar European wild mammals are red squirrels (Sciurus vulgaris). *Utilizing forest trees for homes, food, and routes of escape from predators, they have powerful claws on their front and hind limbs to facilitate rapid, sure-footed movement.* Above. *In a French forest, baby red squirrels sleep in a nest.* Opposite. *An adult squirrel is sheltered by a tree trunk.*

A Tough Adversary: The Hedgehog

Widely admired because it helps reduce harmful insects, worms, snails, and slugs, the hedgehog *(Erinaceus europaeus)* is one of Europe's best-known mammals. It has even been introduced into new regions, as on the Frisian and Baltic islands. Popular fancy has long elevated the hedgehog to almost mythological status. One reason is that the animal has 8,000 spines growing closely together on the creature's back. When the hedgehog is threatened, it rolls its body into a ball-like shape and causes the spines to become rigid, creating one of nature's best defensive mechanisms. The animal is literally stuffed inside a spiny sack, with head and feet drawn inside by special muscles. There have long been claims that the hedgehog, merely by rolling over on fallen apples, could impale them on its spines but this is pure fiction. Its sense of

Hedgehogs range throughout
Europe and into Africa and Asia.
Their 8,000 sharp spines are shed
and replaced individually. These
mammals consume so many
invertebrate "pests" that they have
been introduced for insect control
in agricultural areas.
Above. A hedgehog (Erinaceus
europaeus) in a French forest.
Right. Wetting its spines with
whitish saliva is believed to be
either a cleansing act or a defense
mechanism of the hedgehog.

smell serves importantly in finding food; hence the animal constantly sniffs the air, and when its nose tells it that something edible is nearby, the sniffing increases to a loud snort. This sound has given rise to superstitious beliefs concerning ghosts or evil spirits lurking.

Occasionally, one sees a hedgehog anointing the tips of its spines with saliva. Some investigators have found that contact with toads which have toxic substances in their skins may trigger this activity—the hedgehog transferring such substances to its spines presumably to inflict pain on potential attackers. But the hedgehog does not have much need for elaborate defense systems, considering the fundamental protection afforded by its spines, which helps account for its ability to survive and hence its wide distribution. Its easy adaptation to sand dune habitats, woods, marshes, meadows, and mountains also assures an extended range. Although individual animals may never roam more than a few hundred meters from home territory, eight species of hedgehogs in all are found from the British Isles across Europe and Russia to Korea and eastern China, as well as in Asia Minor and in Africa south from Morocco to Angola.

Life in the Soil
Still, few phenomena are as fascinating as those which take place on the forest floor or in the soil, usually unseen by human eyes. The work of small plants and animals, especially those microscopic in size, has a profound effect on the forest, principally by enriching and aerating the soil in which trees, shrubs, and herbs grow.

Decomposition of the dead and dying remains of green plants is accomplished through the action of many organisms, including the abundant soil protozoa. Decomposition—or, more precisely, the separation of plant or animal material into its constituents—starts even before the trees or their limbs or leaves have fallen. Minerals produced from decaying litter and other debris that falls to the forest floor, or from rotting stumps and logs, are taken up by roots; thus, the growth of trees depends on a healthy process of decomposition.

In northern or upland forests cold temperatures slow this process, but decay proceeds nonetheless. In milder temperate forests, soaked with rain more than dusted by snow, decomposition occurs more actively, and the nutrients are distributed more rapidly and widely in soils. Fungi attack the easily decomposable sugars in leaves, and bacteria help break down cellulose. Soil-dwelling animals recycle the litter by chemical or mechanical means, or a combination of both.

The earthworm *(Lumbricus terrestris)* presents one of the most vivid and easily observed examples of the recycling properties of decomposition. In northern temperate forests the earthworm breaks up soil particles mechanically and passes them through its body as food. Altogether, the sheer mass of other soil animals doing much the same thing is striking. After microscopic protozoa, roundworms (nematodes) are the most widely distributed soil organisms. A cubic centimeter of soil may contain between 1,000 and 10,000 of them.

Belonging to an enormous plant group known as fungi, mushrooms and puffballs are primitive plants occurring throughout the forest, where they grow on live or dead trees, among roots or on tree trunks, leaves, fruits, or seeds. Many forest animals thrive on them, and squirrels sometimes store mushrooms in tree hollows to provide food, in dried form, during the winter. Top to bottom: honey fungus (Armillariella mellea); shaggy inky caps (Coprinus comatus); fly agaric (Amanita muscaria); puffball (Lycoperdon perlatum).

Among the insects most helpful to man is the honeybee (Apis mellifera), *which produces energy-rich honey and pollinates flowers of many plants.*
Top. *A honeybee nest in a tree hollow in an English woods is built up of round waxy combs hanging vertically.*
Above. *When a foraging honeybee returns to the hive, it transfers nectar to other bees which roll it about on their sticky tongues, to help evaporate the water and concentrate the sugars.*

Larger creatures such as snails, mites, and springtails (Collembola)—the latter a very abundant and widespread order of primitive arthropods—also help to replenish soil minerals. How can we know what quantities of forest litter these animals process? It has been estimated that all soil organisms, including such vertebrate subterranean dwellers as moles, voles, and shrews, consume about 40 percent of the forest debris that falls each year. This continuing process reminds us of the fundamental resource that supports each annual rebirth of forest growth.

The Society of Bees

Few invertebrates are as useful to the forest, or produce so copious a quantity of food—honey—edible to human beings, as bees. They are vital to the fertilization of certain plants, for as they move from flower to flower seeking nectar they transfer pollen; this is a necessary part of the reproductive process where flowers have adapted to the pollinating activity of bees.

Most remarkable about bees is their behavior, the intricacies of which are beginning to be understood, thanks to the researches of such eminent scientists as Karl von Frisch. Evidence suggests that bees have been living in colonies and working together for millions of years, evolving into extremely advanced social insects. Some serve as queens of the hive, some as drones to fertilize the queen, and others as workers whose duties include collecting pollen, making wax, and feeding the larvae. The site of the natural hive is often a hollow tree trunk or the underside of a cliff overhang. Since the flowers in bloom at one time may differ in distance and direction from those in bloom at another time, we might conclude that each bee had to waste a good deal of time seeking new flowering trees or meadows as the blossoms successively appeared. But this is not so, for communication among these tiny animals is perhaps the most remarkable result of their long evolution.

In gathering pollen, a foraging bee collects from only a single kind of flower during one day if the supply of that flower lasts. By actions that have been referred to as "dances," foraging bees inform their hive mates of new sources of pollen. As near as can be determined, these "dances" somehow inform other bees how far to fly and in which direction. This mysterious communication is combined with constant exchange of food and glandular secretions, by means of which the inhabitants convey the needs of the hive—when to swarm, when to start a new colony, and so on.

In his researches on the behavior patterns and sensory capacities of bees, Von Frisch gives us some extraordinary insights into the life of these insects. "If bees have discovered a good feeding place," he wrote, "they announce the fact in the hive by means of certain dances performed on the honeycombs. The other bees not only learn that there is food available, but they are also informed in which flowers it is to be found. They obtain this information from the scent of the flowers which adheres to the bodies of the dancing bees."

Danube Wildlife

The Danube River flows some 220 kilometers east of Vienna before making a sharp turn southward toward Budapest. Within the fold of that turn is the unusual forest of Pilis, a relatively wild deciduous woodland maintained by the Hungarian government. This forest follows the west bank of the Danube for about 70 kilometers and ranges from 10 to 30 kilometers wide. The underlying strata are dolomite, limestone, and volcanic rock, resulting in a variety of soil habitats from which a varied forest has sprung. Characteristic of the flora are the downy oak *(Quercus pubescens)*, turkey oak *(Q. cerris)*, and hornbeam *(Carpinus betulus)*. The forest meanders over low mountains and into ravines, where one finds common ash, maple, and elm; alders grow along stream valleys. Most impressive are the beeches, with their crowns closely linked, their tall bluish-gray trunks rising like immense columns in the dim light. A small copse of late-flowering chestnut trees may be found in the castle garden at the village of Visegrád.

The surrounding forest supports nearly all the characteristic species of Hungarian birds, including such birds of prey as lanners *(Falco biarmicus)*, peregrine falcons *(Falco peregrinus)*, kites *(Milvus milvus)*, and imperial eagles *(Aquila heliaca)*. There are thousands of invertebrates as well as numerous forest mammals, such as weasels, squirrels, badgers, ermines, foxes, and wildcats, many of which prowl at night. Larger animals include deer, mouflons *(Ovis musimon)*, and wild boar *(Sus scrofa)*. Curiously, the boar had survived longer in Hungary than elsewhere in Europe because parts of Hungary were occupied for more than 200 years by the Turks, who did not touch boar meat because of Muslim dietary laws. The Turks also forbade anyone to hunt boars with weapons, including knives and spears; consequently the animals proliferated until World War II. After the war landless peasants rushed to divide up the huge private estates, cut valuable forests for lumber, and slaughtered deer and boar for food. By 1946 fewer than a hundred wild boars remained; at that point the Hungarian government forbade all hunting of game animals, while gradually developing areas like the Pilis Forest as wildlife preserves and bird sanctuaries. Today, by far the largest numbers of wild boars live in Hungarian wildlife preserves.

Boars may seem awkward, but they are sure-footed and quick on land and fine swimmers. They are not to be trifled with: the hogs have four continuously growing tusks, two in each jaw. Usually they do not attack, but when molested they may become savage—a situation all the more perilous if the animals are traveling in a large band. Gregarious by nature, they are most active in the evening and early morning hours. Intent on roots, nuts, plant stems, insect larvae, and carrion, they may seem oblivious to the world at large. But that diet nourishes them to weights of as much as 200 kilograms, so that any intruder, human or animal, in their domain would be well advised to keep a respectful distance.

From the city of Szekszárd to the Yugoslav border, a distance of 80 kilometers, the Danube often goes on a

64–65. Wild boars (Sus scrofa) with their young forage in a German forest. Boars are prolific, and may have two litters a year with a dozen young; their life span is 15 to 20 years. Boars have long been hunted throughout Europe, not only as a source of food but to prevent their destruction of crops.

rampage in the spring, flooding large areas of dense woodlands. A "flood forest" is created, with numerous small marshy lakes and ponds that remain through the summer. Because moist, warm Mediterranean air reaches this area, a distinctive microclimate is formed that results in a lush vegetation referred to as "the Hungarian jungle."

Carpets of Crocus: The Carpathians

Among the most dramatic forests in eastern Europe are those on the Carpathian ranges, which curve for more than 1,600 kilometers from the Czechoslovak-Polish border through a corner of European Russia and down into Romania. The Carpathians are as long as the Alps but not as high; since they nowhere rise more than 3,000 meters above sea level, they are more widely clothed in woods and meadows. In the High Tatra Mountains is a tiered arrangement of montane vegetation: fir and beech mixed with pine, larch, and sycamore maple up to 1,250 meters; spruce from there to 1,550 meters; and mountain pine shrubs to 1,990 meters. Red deer, roe deer, and boars inhabit these forests, along with numerous other mammals. One estimate of the more important types included 30 wolves, 230 bears, 30 lynxes, and 100 wildcats, with ermines, weasels, badgers, otters, foxes, and martens also seen.

A favorite of naturalists and hunters is the lynx *(Lynx lynx)*, an aggressive wildcat that stalks and pounces on whatever it can find, from prey as small as mice to animals as large as deer. If the prey is too large, however, the lynx may settle for young, old, or diseased individuals. And if the lynx tangles with a wolf, it may not come out of that encounter alive. Intensely territorial, the lynx eliminates what interlopers it can, then fights other males of its own species during the spring mating season.

The air of the Carpathians sometimes carries the melodies of pygmy owls or the songs of red-breasted flycatchers *(Muscicapa parva)*. For pure luxuriance, one can scarcely surpass the dense carpets of flowering crocus *(Crocus albiflorus)* found here. Other Carpathian ranges, the Pieniny and Bucegi, contain some extraordinary collections of plants and animals, including one area where 1,800 species of moths and butterflies have been recorded.

Azaleas in the Spring: The Caucasus

The Caucasus, a system of folded, glacier-coated mountains extending east from the Black Sea to the Caspian for more than 1,100 kilometers, separate Europe from Asia and rise to the perpetually snow-capped prominence of Mount Elborus, at 5,629 meters the highest mountain in Europe. The upper meadows and the thickets of rhododendron scattered through forest zones of birch and maple, then oak at the next level, are home terrain for the Caucasian black grouse *(Lyrurus mlokosiewiczi)*, a handsome bird lacking the lyre-shaped tail of the European species. But, as in all such alpine regions, we do not see half the organisms that live in the ground, burrow beneath the surface (or beneath the ice and snow in colder seasons), sift the soil, and survive with ease. These include snow voles *(Microtus nivalis)*, forest dormice *(Dryomys nitedula)*,

and pygmy susliks *(Citellus pygmaeus)*. If they are not specially adapted to withstand icy temperatures, they at least know how to retreat from the cold and stay where it is relatively warm.

One of the great delights of this region is the abundance of azaleas (as along the Blue Ridge Mountains of eastern North America), whose large orange aromatic flowers burst forth at elevations from sea level to 2,100 meters. As a colorful understory in oak-beech forests, these masses of azaleas provide one of the principal springtime dramas of the Caucasus.

Passing through cutover forests in the lower parts of the Caucasus is not particularly rewarding; and if we are not careful, we shall be scratched by sharp hawthorn or blackberry patches, or entangled in wild grape and clematis vines. But if we seek some specific kind of forest, it is likely to be found in one part or another of these diversified mountains. The slope exposure makes a difference, as does the wind—which in places blows more than 20 meters per second, lasts for days, keeps trees from taking root, and accounts for only a limited meadow growth. In some parts of the Caucasus, more than 2,000 millimeters of rain wash the woods each year, and the vegetation takes on a subtropical aspect. The lower and warmer locales have Mediterranean-like environments, where forests of juniper and Aleppo pine flourish. If it is variety of forests that one seeks, from near-tropical to alpine, the Caucasus is the place to find it.

The Great Survivor: Red Fox

When we speak of variety of habitats, there is one animal group that can be found almost everywhere in Europe's— and indeed, the world's—forests. High and low, cold and hot, wet or dry, wild or domestic: the environments may vary widely, but the fox is likely to be there. From Arctic to tropics, one may find red foxes *(Vulpes vulpes)* and gray foxes *(Urocyon cinereoargenteus)*, along with numerous other species.

Whether this high survival rate is due to any sly or crafty behavior that folklore tells us the fox has can be left to researchers in canid ethology. The fact is that the fox has become especially adapted to habitats radically altered by man. The question now is: Can man become adapted to the fox? One answer is, not where rabies is concerned. Let one case or one rumor of suspected rabies be announced, and every fox, rabid or not, is thereafter shot on sight— nearly 200,000 a year in Germany alone, and many more elsewhere. Or let one fox raid a chicken farm, and the reputation of all suffers.

These conflicts, whether based on truth or fallacy, have so far been met and overcome in almost epic proportions by the red fox; but its success may not last much longer, for through heavy hunting pressure and den destruction the fox has been exterminated in some places and badly decimated elsewhere. Although its natural enemies such as the white-tailed sea eagle *(Haliaëtus albicilla)* have also diminished, that is scant comfort in the face of the onslaught by men.

Accordingly, the fox has had to make an evolutionary

Placed near the bottom of many food pyramids, mice and voles form vital food sources for certain birds, reptiles, and mammals of the forest community. When mice and voles become scarce, the animals dependent on them for food also diminish in numbers.
66 top. A dormouse (Muscardinus avellanarius), found in thickets and forests throughout Europe and into Asia Minor and the USSR.
Center. The birch mouse (Sicista betulina), seen perched on a coniferous branch, lives mainly in deciduous woods and thickets, where it feeds at night on seeds, berries, and insects.
Bottom. A bank vole (Clethrionomys glareolus) observed in an Austrian forest. These rodents are active at all times of day and night and in all seasons.

Above. A dormouse (Glis glis) with its young. This animal takes on a dormant condition for long periods in winter, but occasionally wakes to feed on stored nuts and seeds.

changeover from natural to unnatural habitats, except where parks and reserves offer safety. In these recovering natural areas, we get some idea of the astonishing adaptive powers of this animal. For one, it eats almost anything, though its preferred foods are mice and voles; if these are plentiful, the fox will contentedly remain within a home territory of a few square kilometers. Berries, fish, insects, and carrion may be eaten as well, but if the fox is really hungry and the opportunity arises, it will pursue larger game and eat just about anything: young boar, fawns, hares, pheasants, grouse, domestic fowl.

The fox digs its own den, unless it finds a convenient badger den; if so, it may move right in and establish a place for itself and its family in harmony (more or less) with the other occupants. Old dens may have more than a dozen entrances and on occasion serve the forest community by providing shelter for other animals such as wildcats, rabbits, and small owls.

Without the fox—who knows? There is an old German proverb about a fox crossing newly formed ice on streams and ponds in winter: *Trägt's den Fuchs, so trägt's den Jäger* ("If it holds the fox, it will hold the hunter"). That proverb may also carry a contemporary lesson which can be applied to the world environment: *If the fox goes, so will man.*

In the Colder Regions

Scandinavians, particularly in the last half-century, became acutely conscious of their natural treasures—wild forests and populations of such conspicuous animals as deer, wolf, wolverine, elk, and bear.

For the energetic hiker, there are some notable sights in back-country Sweden. Nearly half of Muddus National Park, in the coniferous belt of northern Norrland, is covered with bogs that in summer are white with cotton grass (which is neither cotton nor grass, but sedge). In the center of dense woods is a sanctuary for breeding birds, notably the whooper swan *(Cygnus cygnus)*, and one can hear the wailing cry of black-throated divers *(Colymbus arcticus)* or watch ospreys *(Pandion haliaëtus)* fly over and ruffs *(Philomachus pugnax)* strut and spar.

The farther north we go, the larger the fur-bearing mammals become. Siberian roe deer *(C. capreolus pygarus)*, for example, are larger than the European variety and have much stronger antlers.

By far the largest single forest zone of the USSR is the taiga coniferous forest, a natural consequence of the country's location almost entirely north of the 40th parallel. Areas of mixed forest and forested steppes are relatively small. ✓

The Soviet taiga is not a simple "land of little sticks," as the name is supposed to signify. There are, in fact, spruce taiga and pine taiga, in addition to transitional areas between fir and pine.

The taiga once teemed with animal life. It was the home of large numbers of bear, reindeer, roebuck, and lynx, but many animals have been decimated or eliminated by man and are now to be found principally in remote or protected areas. Chipmunks and flying squirrels are typical of the taiga. Some birds such as the hazel grouse *(Tetrastes bonasia)* and willow grouse *(Lagopus lagopus)* nest

71. *Now much restricted from its former range, the capercaillie* (Tetrao urogallus) *lives only in large stands of spruce and pine with an understory of berry-bearing shrubs. Seen here, in a Norwegian woodland, a cock displays in order to gather a harem of hens; the male caper-caillie will fight vigorously to maintain its territory.*
72–73. *The wisent, or European bison* (Bison bonasus), *survives in this primeval forest of Białowiecźa in Poland. Once extinct in the wild, wisents were sheltered and bred in zoos and animal preserves, then released into natural forests again. Wisent calves are protected by the bull or the cow or, when necessary, by the entire herd.*

and live there year-round; they have become so well adapted to the available food and to conditions there in general that they simply do not leave.

The Capercaillie and Its Mating Dance

One of the most interesting birds of these spruce and pine forests of Europe is the capercaillie *(Tetrao urogallus)*. The largest of the grouse, the male capercaillie has a massive body almost a meter in length. A darkish bird, it is readily identified by its size, the red wattles over its eyes, and a rough beard of dark feathers. Since it is a wary creature and keeps to deep woods, it is hard to detect. It feeds on pine needles and buds.

The most fascinating habit of the capercaillie is the male courtship dance, performed in late winter or very early spring at a regular meeting ground, usually with several males participating. Just before dawn, on a tree limb or on the ground, the male bird goes into a kind of trance and utters a series of calls ending in a hissing sound. On a limb he does a sidewise dance and finally points his bill skyward, stretches his wings downward, and spreads his tail like a fan—truly the cock of the walk. If the performance is on the ground, he will strut about, threaten other males, and attack them. After this curious ritual the males and the females mate.

The Massive Wisent

Of all the changes throughout the centuries, expanding human settlement and extensive hunting have done the most damage to the European bison or wisent *(Bison bonasus)*. Once found across Europe and Asia, from the Atlantic to the Pacific oceans, these beasts eventually retreated to a last stronghold, the forests of Białowieża in Poland. There the last wild bison was shot by a poacher on February 9, 1921.

Fortunately, 56 specimens had survived in zoos and private game reserves, and a concentrated, seemingly hopeless effort began to restore the animal in the wild. Believed by some scientists to have evolved from *Bison sivalensis*, which formerly lived in India, the animal is known from many cave drawings of the glacial period. Millennia ago, some herds evidently migrated from Asia across a northern Pacific land bridge and then evolved into the North American bison.

That is a notable range, and the European bison, called by some the most majestic and powerful of all European animals, was too important a natural legacy to lose. After the formation of the International Association for the Preservation of the European bison in 1923, careful breeding of captive animals in Poland, Germany, the Soviet Union, and elsewhere led, in 1956, to the release of a small herd of wisents in Poland's Białowieża National Park. The experiment proved successful, for by 1963 the population of the free herd had increased to 57 animals, and another small herd was released in the adjoining Soviet portion of the forest. By 1967 the world total of European bison was 860—more than fifteen times the remaining stock in captivity in 1923 and enough to encourage real hope of survival.

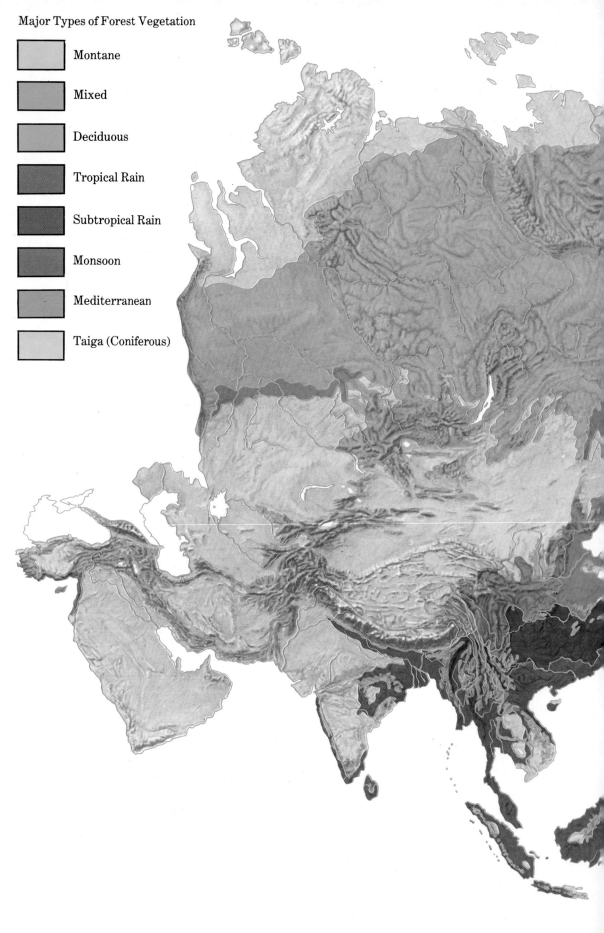

Major Types of Forest Vegetation

- Montane
- Mixed
- Deciduous
- Tropical Rain
- Subtropical Rain
- Monsoon
- Mediterranean
- Taiga (Coniferous)

Asia's Varied Forests

The woodlands of Asia embrace the Earth's major vegetational zones, from tropical to tundra, from montane to monsoon. Of special interest are the vast rain forests which lie principally in Burma, Thailand, Malaysia, the Philippines, and the islands of Indonesia, covering an area of 240 million hectares that makes Asia's forests second only to those of South America in extent. There are also widely separated patches of either coniferous or deciduous forests in eastern China, Japan, Korea, and parts of the USSR. To the north, the conifers of the taiga reach across an immense landscape from the Urals to the Pacific. These rich woodland areas almost mask the fact that much of Central Asia consists of nonforested lands, deserts, steppes, and the enormous mountain ranges of the Himalayas.

Nepal's Rich Monsoon Forest

Our tour of Asian forests began in one of the most remarkable wild regions remaining on the Indian subcontinent: Chitwan National Park in southern Nepal. We disembarked from our plane at Meghauli, climbed aboard our elephants, and set off for the Rapti River. It took about fifteen minutes to get there, splash across the ford, and enter the park. Almost at once we were engulfed in the remnant of a subtropical monsoon forest that once extended thousands of miles east and west of us. Barely visible far to the north rose the white ramparts of the Himalaya Mountains. Here in the lowlands of Nepal, at an elevation of only about 100 meters above sea level, we could sense the richness and humidity of Asia's great tropical forests to the east and south.

Chitwan was established as a wildlife sanctuary in 1962 to preserve the one-horned rhinoceros and then, nine years later, became Nepal's first national park. Covering nearly a thousand square kilometers, it protects part of an immense stand of sal *(Shorea robusta)* in a more or less original condition. This is not a tropical rain forest, for the trees are not evergreen but deciduous, losing their leaves in the dry season. Furthermore, this forest is more open and its canopy less dense than one finds in a true tropical rain forest. A monsoon forest, where nearly all rain falls during the monsoon period, Chitwan endures an intermittently stormy season between June and September. At that time the vegetation bursts forth, as well it should: the rain in July alone amounts to nearly 700 millimeters, and the annual total is 2,156 millimeters.

The Solitary Tiger

At Tiger Tops, a jungle lodge in the center of Chitwan National Park, on a recent trip we were hoping to see tigers *(Panthera tigris)*, about thirty-five of which inhabit the park. But we did not see any. Tigers are shy, nocturnal, solitary, and unpredictable. Considering their precipitous decline in numbers, it is almost a miracle that any remain. Fifty years ago about 100,000 tigers of all species existed in the world, with perhaps 40,000 found in India alone; now there are only 5,000 worldwide. About 2,000 of the Indian variety of tiger remain in India, Nepal, Bangladesh, and Bhutan.

Life for this largest of cats is not easy, despite its position at the apex of the pyramid of forest life. To bring down some passing morsel of food, the tiger must do more than merely lift a paw and claw now and then. A tiger may have to hunt its prey for days, simply because most of its associated species have developed a capacity to avoid the much-feared tigers. A cat so huge must therefore use subterfuge, stealth, and surprise. This is hard work, and even the most cunning tiger may go hungry for several days. If the tiger misses its mark on its initial surprise leap, it will pursue prey for perhaps 200 meters, then give up. Though of large size, weighing as much as 272 kilograms, the tiger walks the forest quietly, swims with ease, and can cover 6 meters in a single bound. When successful in its hunt, it tears the prey apart and eats its fill, then covers the remains, if any, with grass, twigs and leaves, and lies down to rest.

As a rule each tiger leads a solitary life, except at mating time and while the young are with their mother. Each adult has a den, or perhaps several dens, within its territory. These may not amount to much—a cave or crevice, a space under a fallen tree, a dense thicket, perhaps with the floor covered with dry grass and foliage. In such retreats the tiger sleeps and relaxes or, if female, sets up its nursery. All this represents a natural style of life that governments are attempting to restore. Just as once the most dramatic efforts were made to kill the tiger, so now they are made to save it. India not only has banned tiger hunting, but in 1973 launched Project Tiger, designed to set up and administer nine reserves in different tiger habitats to ensure the survival and maintenance of tigers in India.

As the chief naturalist at Tiger Tops said recently: "Let's face it. Unless human population expansion is controlled, then it's all over for wildlife in a matter of time."

Chitwan by Elephant: From Barking Deer to Wailing Jackals

Even though we failed to observe tigers during our stay, we were fortunate enough to glimpse many other species of wildlife inhabiting this vast park. Lurching on the backs of native elephants, a natural means of travel through these forests, we caught sight of wild boars *(Sus scrofa)* moving along the banks of a small stream. Ruddy shelducks *(Tadorna ferruginea)* flew overhead. In the distance we caught a flash of iridescent blue above the trees and watched a pair of peafowl *(Pavo cristatus)* float lazily down and land at the edge of a stream. Searching for seeds and other vegetable matter, they wandered in and out among the patches of riverside grass until they disappeared downstream. Had we been there at the proper time and searched out their mating grounds in clearings in the sal forest, we might have observed the celebrated displays of the male peafowl preening before their harems of two to five hens.

A flock of red jungle fowl *(Gallus gallus)* flew through the forest and out into a forb-filled glade, commencing to peck like chickens in a village. Resembling English game bantams, they seemed strangely incongruous in this wild setting. But not so, for this has been their home for millennia. These jungle fowl are the chief wild ancestors of

all domestic poultry. Today their distribution coincides almost exactly with that of the sal tree.

Vociferous jungle fowl often accompany gaurs *(Bos gaurus)*, largest of the world's wild cattle. As the jungle fowl feed on insects, in association with grazing gaurs, their behavior alerts the other animals to approaching danger.

Gaurs, which weigh up to 1,000 kilograms and reach over 200 centimeters maximum shoulder height, prefer grazing in meadows with adjacent hilly forests and streams for drinking and bathing. Though shy, they move about in herds of as many as twenty, alert to danger (tigers, according to one estimate, take half their young), and when showing belligerence, approach their enemies broadside rather than head-on. Their formidable horns are much respected by tigers.

At one point in our foray, the mahouts suddenly shouted instructions to the elephants, which turned and lumbered rapidly into the dark sal forest, where the sharp eyes of the drivers had detected deer so perfectly camouflaged that the ordinary visitor might not have seen them. In Chitwan live tiny barking deer *(Muntiacus muntjak)*, with deep chestnut-red coats; hog deer *(Axis porcinus)*, somewhat similar to barking deer: chitals *(Axis axis)*, seen in herds of up to 60; and large (318 kilograms) sambar deer *(Cervus unicolor)*. These animals have a predilection for young shoots, just as rhinos have; but, unlike rhinos, they are themselves attractive prey for Bengal tigers.

Much more conspicuous than the deer, or at least more noisy, were the common langurs *(Prebytis entellus)*, arboreal monkeys that chattered as they leaped among sal branches—their lives dependent upon avoiding the equally arboreal leopards. Rhesus monkeys *(Macaca mulata) also inhabit the Chitwan park, but mostly in the Bombax* forests along its major rivers.

It takes a little searching to find the sloth bear *(Melursus ursinus)*, which is neither a sloth nor slothful; it is one of the most industrious animals in the forest. A shaggy black bear with a white V on its chest, it has a long snout and claws 10 centimeters long and weighs approximately 90 kilograms. The sloth bear eats fruit, insects, and honey, as well as termites from mounds ripped open with its claws.

If one stays long enough in Chitwan, one may see some of its other abundant wildlife: wild dogs, small cats, mongooses, flying squirrels, porcupines, crocodiles, monitor lizards, Indian pythons *(Python molurus)*, king cobras *(Naja elaps)*, and more than 300 species of birds. Every visitor takes away some indelible impressions: ours was the mournful wail of jackals at night in the sal forest.

India's Vanishing Wildlife

The overwhelming human population on the subcontinent of India, as well as the spread of fields and farms and the prevalence of hunger, has reduced its forests by 75 percent; correspondingly, its wildlife has diminished alarmingly. Although a remarkable variety of ecosystems remains, from rain forest to desert to mangrove forest, the great abundance of wildlife that may linger in the popular mind is now a fact of the past. The lion *(Panthera leo)* has been

80–81. *A solitary hog deer* (Axis porcinus) *with a myna on its back pauses after feeding in Chitwan National Park, Nepal. Generally found in groups of two or three, hog deer seldom venture into deep forests but prefer light woodland growth, where they subsist on grasses and fallen flowers and fruits.*

Right. *An Indian python* (Python molurus) *in Chitwan devours a hog deer, which had probably stopped to drink in the snake's watery realm. The snake captures food by darting quickly toward the prey and seizing it with its long sharp teeth. The predator then wraps itself round the victim and strangles it or ruptures its vital blood vessels. After swallowing such a meal, a python may not eat again for weeks.*

reduced to a few survivors in the Gir Forest of Gujarat. Once-great herds of barasingha *(Cervus duvauceli)*, which resemble the European red deer *(C. elaphus)*, are no more.

Where tropical forests have remained relatively undisturbed, some trees reach heights of nearly 50 meters, forming an umbrella-like canopy. Shrubs are scarce underneath, so the forest somewhat resembles those of Africa and the Amazon River basin of South America. Elsewhere, forests range from moist to dry, and in them we can note, with some comfort, that wildlife still includes, for example, small groups of blackbuck *(Antilope cervicapra)*, blue bulls *(Boselaphus tragocamelus)*, and hornbills (Bucedrotidae). The latter may be seen in noisy flocks flying from tree to tree in search of fruit. Some have a roaring call that sounds like laughter; some are so large that their wingbeats sound like a puffing steam engine.

Hornbills nest in the cavities of trees, but the process differs considerably from that of most other birds. When the female has entered the nest and laid her eggs, the male seals the nest entrance with clay leaving only a small opening. The female dwells in this natural prison until the eggs are hatched and the young are partially grown. The male must, of course, feed the nest's occupants until they are ready to leave, at which time he breaks the seal and liberates them. This may seem like a bizarre and unnatural system, but if one recalls the hazards of Asian forests—the prevalence of egg-loving predators, for example—it becomes a highly practical way of assuring that the female never leaves the nest unattended, so that the eggs and young are constantly protected. This nesting behavior is not unique with hornbills; female Magellanic penguins of South America also guard their nests without interruption until the young are ready to leave.

Monkeys and Tigers in the Snow

Across Asia the pattern is of more or less diminishing forests. The Siberian taiga landscape, however, includes forests of birches and aspen, which give way to more durable firs, spruces, larches, and pines. In these environs a number of fascinating animals survive. The northernmost of all monkeys, the Japanese macaque *(Macaca fuscata)*, has become adapted to cold; yet the sight of it cavorting in deep snow still strikes the observer as incongruous.

The more southerly deciduous forests of the Far East benefit from the widespread monsoon. Conifers add some variety to them, but birches, maples, elms, and lindens predominate. Beneath these prevalent varieties grow lesser trees typical of other deciduous forests: hornbeam, hawthorn, and ash. The autumn color is magnificent. Deer inhabit these woods; so do tigers *(Panthera tigris longipilis)*, another species seemingly out of place as they wade through winter snow. This northern tiger is larger than its more tropical and subtropical counterparts, but much rarer; the Soviet government estimates their population to be a scant 125.

These temperate deciduous forests once extended much farther south into China, but no longer. Expanses of oak do

84 *top. Tropical Indian and Asian forests provide habitats for various genera of hornbills (Bucerotidae), all of which have large toucan-like beaks used in eating fruits and nuts. The nasal trumpeting call of the great hornbill* (Buceros bicornis), *shown here, is one of the most common sounds in both Asian and African forests.*
Bottom. *Many forest predators, such as the green tree snake* (Ahaetulla nasuta) *of southern India, have become important links in the woodland chain of life. The prey they capture after lying in wait among branches and leaves may, in turn, have fed directly upon forest vegetation.*

exist there, but the forests of mainland China, together with their wildlife, have been largely decimated during centuries of cultivation, warfare, and settlement. Only much farther south, into the tropics, do the woods recover some luxuriance as they continue into the Indo-Malaysian region.

The Woods of Tropical Asia

The equatorial rain forests of Southeast Asia, Africa, and South America may look alike to a casual visitor—and, indeed, in profile they might be called "typical rain forest." But the species of plants and animals found in the three areas are vastly different; few genera recur, and some are unique. Such diversity is a principal fascination of the world's rain forests. In the Philippines, for example, lives the endangered monkey-eating eagle *(Pithecophaga jefferyi)*, which occurs only on the island of Mindanao, where only some forty pairs remain. This wild-looking bird with flared head feathers lives in the tops of giant kapok trees *(Ceiba pentandra)* on remote slopes in some of the most inaccessible rain forest in tropical Asia. Belying its name, it does not feed exclusively on monkeys but may capture birds, domestic animals, or related organisms that come within easy range. The problem affecting its survival is limited reproduction: these eagles usually raise only one offspring a year.

There are abundant trees—hundreds of species each of dipterocarps, including the giant *Shorea* that tower to 60 meters above the forest floor and the aromatic *Eugenia, Ficus, Rhododendron,* and *Vaccinium.* All this flora, growing at different levels from floor to canopy, and sometimes plant on plant on plant, constitutes some of the world's richest vegetation. Climbing palms, rattans, heaths, air plants, bamboo thickets, giant ginger, and even a few pines are all in evidence.

The largest single family is that of the orchids; Indochinese rain forests alone nourish some 5,000 species. Not all have the familiar large conspicuous flowers; the blooms of some are dull and unobtrusive. Showy wild orchids near human settlements have long disappeared, owing to successive generations of overzealous orchid hunters. In Java and the Sunda Islands, however, there are still trees covered with the white flowers of *Dendrobium,* a typical Asian orchid. In northern Bengal and Assam, orchids abound in the forests of both plains and mountains. Most highly prized is one of the rare perfumed orchids *(Aerides odoratum)*, which give off a faint honey-like fragrance.

Rain Forest Creatures: Minute and Massive

With such an extraordinary vegetative base, these natural forests abound in wildlife. During a walk through this region, one could encounter three times as many types of animal life as in a warm temperate forest. By looking carefully, a visitor might discern dozens of species of snakes and lizards and countless kinds of insects. If you were to spend years here, as naturalist Alfred Russel Wallace did between 1854 and 1862, you might approach his listing of thousands of species of birds and other native life forms.

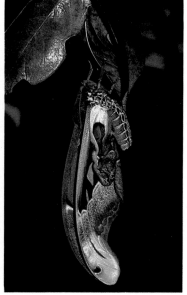

You would have a great deal of trouble locating a few mammals, however, because they are nearly extinct. If you went to the Udjung Kulon Reserve on Java and were allowed to enter, you might glimpse one of the extremely rare Javan rhinoceroses (*Rhinoceros sondaicus*), which now number about two dozen individuals. Their home has been sealed off from human intruders; thus protected, they are free to munch grass and reeds, roll in the mud, and perhaps increase in numbers. That is the best man can do for this dwindling species.

More common are the strictly arboreal species: squirrels, monkeys, and apes such as gibbons—tailless, acrobatic primates with greatly elongated forearms and hands. So agile are they with their strong limbs that they have become the fastest of the primates, hurtling themselves through the forest with a speed approaching that of flight. They swing into space on long dangling lianas, leap as far as 12 meters, and may even catch birds in flight. They fill the dense, foggy forests with resonant, booming songs at sunrise, so earning a reputation as the "howler monkeys" of the Old World. A sound-amplifying throat sac permits a flood of tonal reverberations that drowns out all other noises of the forest. This is no ode to sunrise or joyous wake-up call; the gibbon is apparently proclaiming its home territory before going out to feed on leaves, fruit, buds, ants, snails, and whatever else the forest provides.

The Perilous World of the Orangutan

The endangered Malaysian orangutan (*Pongo pygmaeus*) is the most arboreal of the great apes. Because it evolved in trees and spends so much time swinging from limb to limb or grasping trunks, the orang's arm spread is wide—some 2.25 meters—and its arms are more strongly developed than those of other great apes. On their travels, orangutans chew bark and rotting wood, pick and consume mushrooms, search for fruit, capture insects and lizards, munch on leaves and young shoots, and, with a little luck, take eggs from any birds' nests they happen to find. To scout for food and enemies, they often climb a high treetop, mountain ridge, or rock. They make nests high in the trees and bed down to sleep there, though they may suffer from wind, storm, and cold in such elevated locations.

However, these perils are not the principal difficulties endangering them. Their survival problem is inherent in their nature: the females nurse their young for four years, during which no additional offspring are produced. The females may live for thirty years and produce another two or three young—a small number in so perilous a world. In all, presently there are only a few thousand in existence. Despite laws prohibiting orangutan hunting, illegal poaching continues, and the orangs are being pushed into more remote locations and up into the mountains. They cannot climb or remain above 2,000 meters, lest the coolness, together with their lack of protection against it, induce severe colds or worse.

For years Barbara Harrisson, the affable researcher and companion to orangs, has been trying to return orphaned young to a useful and successful wilderness life. With a total of only one or two thousand of these unique animals in

Superior to other forest mammals in agility, gibbons spend much of their lives swinging among branches in graceful hand-over-hand movements called brachiation. The action is so rapid and skillful that gibbons can almost be said to fly.
Top, above. *This whitehanded gibbon* (Hylobates lar) *of Singapore, seen swinging deftly through the trees, possesses the greatly elongated arms and hook-like hands typical of the species.*
89. *A male whitehanded gibbon of Malaysia is seen in its treetop domain. Gibbons generally live in family groups of one male, one or two females, and assorted juveniles.*

The arboreal great ape, the orangutan (Pongo pygmaeus), *travels rapidly from tree to tree with arms of greater strength than those of other great apes. The orangutan sleeps in a treetop platform built of sticks and vines each night.*
Above. *Orangutans swing through the forest on lianas. Because of their diverse dietary needs, these apes roam over large areas of forest in search of food.*
Left. *An orangutan mother with its young, photographed in Sumatra. Young orangs nurse for a prolonged period of three to four years, during which time the mother cannot conceive.*

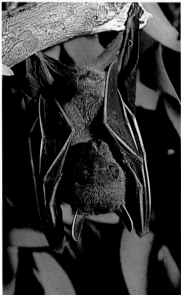

existence, she asks: "Do we still have time? Or is it too late already?"

Wings Over the Forest

Since untold millions of bats inhabit the earth, they exert a considerable influence on woodland ecosystems, especially in the tropics, where they are most abundant. Some eat fruit and spread seeds; some gather nectar and pollen, thereby pollinating flowers just as bees and some birds do. About 20 genera of trees and shrubs have become wholly or partially dependent on bats for pollination of their flowers, including *Bombax* and the African baobab. Other bats capture fish and small vertebrates. But by far the greatest contribution of bats to world ecosystems is keeping the insect hordes in balance. About 70 percent of all bats consume insects—beetles, flies, ants, and virtually any other invertebrate aloft—as their major food.

It was easy for early peoples to think of the bat as a flying, fluttering mouse or a flying fox. But bats are actually far different creatures, the only mammals with functional wings and equipped with a special echo-location ability that allows many to fly in caves and through forests. Because this sonar-type guidance system also leads them to insects and other food, woe unto the bat that loses its hearing. Bat wings, built of the same bones as human hands but greatly elongated, can have an immense spread. The black fruit bat of Indonesia *(Pteropus niger)* has a wingspan of nearly 2 meters. The New Guinea fruit bat *(P. neohibernicus)* may have an even greater wingspan.

Malaysian forests alone have nearly 150 species of bats, ranging from the diminutive *Miniopterus* to large "flying foxes." Some are widespread in distribution, notwith-standing intervening deserts and oceans; the lesser horseshoe bat *(Rhinolophus hipposideros)* belongs to a genus found throughout the Old World. Several "false" vampire bats *(Macroderma* and *Megaderma)* live in Asia; these do not drink blood as the true vampire bats of South America do, but feed on insects and small vertebrates such as frogs and mice. Like hawks, these bats can land on the ground and take off with their prey, often carrying it to a favorite perch, beneath which a pile of bones accumulates from frequent repasts.

In the tropics, a great many bats subsist on fruit, for which they are well adapted. Fruits are crushed in the mouth, with the seeds, skin, and fibrous material compressed into a pellet and then spit out. Their palate has a series of sharp transverse ridges that aid in mashing the fruit. The long tongue of pollen-gathering bats may bear tufts of hair or barbs; when pushed deep into a flower, such a tongue will pick up large quantities of nectar or pollen.

In Asia, as elsewhere, the continuing threats to the environment and to wildlife have had a sobering effect on some governments. As a result, through official efforts, patches of monsoon forest, equatorial woods, and deciduous groves—along with the associated wildlife—are being helped to recover. Even the embattled tiger, king of the Asian woods, may yet survive as a symbol of that continent's varied marvels of nature.

Many kinds of bats have evolved to live in the same area by exploiting different food sources in various parts of the forest canopy.
Top. *The spectacled flying fox* (Pteropus conspicillatus) *is a fruit bat of tropical regions that benefits the forest by seed dispersal as it feeds.*
Above. *A short-nosed fruit bat of Malaysia* (Cynopterus brachyotis).
Opposite. *The Java flying fox is another fruit-eater of the family Pteropodidae. The palate of fruit-eating bats has a series of sharp transverse ridges that aids in crushing the fruit.*
94–95. *Fruit bats are seen hanging from trees in Ceylon. Lacking the ability to echolocate, true fruit bats must rely on their vision and spend their lives in forest trees rather than in caves.*

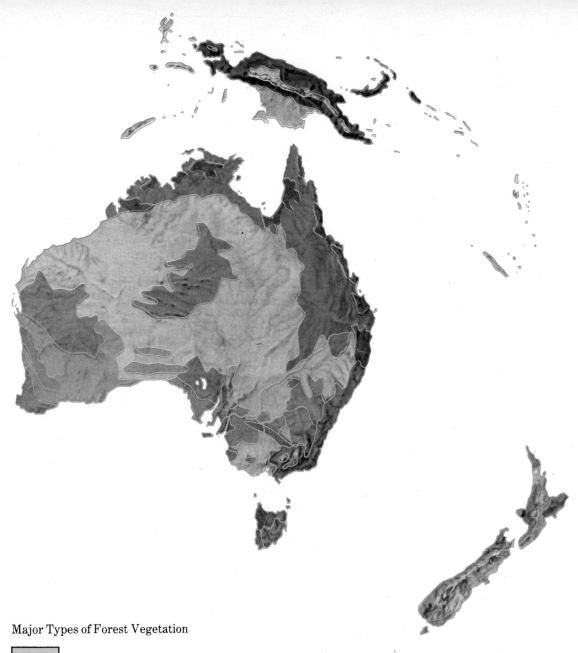

Major Types of Forest Vegetation

 Coniferous

Tropical Rain

Eucalyptus

Mulga

Mallee

Woodlands of Australia and the South Pacific

The natural treasures of Australia surrounded us on every woodland trail, on every ridge, in every ravine; but the abundance of colorful birds was unexpected. This continent is truly a land of birds— especially parrots. Australia's landscape exhibits mostly open desert, harsh sunshine, and treeless plains. Only 5 percent of the continent is forested with trees of commercial size; less than 40 percent is woodland of any kind—hardly sufficient, one would think, for a considerable avifauna. Thickets of brushwood, or "mallees," occur on alkaline soils bordering desert regions. Behind these lie semiarid woodlands mixed with savanna-like meadows and grassland, which is the most widespread vegetative association in Australia. And in patches here and there, especially along the east coast, are rain forests. Some of the best-known trees produce exceptional displays at flowering time, such as the bright red Christmas tree *(Ceratopetalum gummiferum)*; the giant *Banksia*, with its yellow flower spikes over 30 centimeters long; various wattles, including the golden wattle *(Acacia pycnantha)*, Australia's floral emblem; and the flame tree *(Brachychiton acerifolius)*, whose fiery cloud of blossoms grows to 40 meters high.

No trees are more abundant in and characteristic of Australia than the eucalypts, with more than 500 species and hybrids of them making up 95 percent of the forests of Australia. They have adapted to the extremes of climate, wet and dry, hot and cold, as well as to fertile and infertile environments. The world's tallest hardwood, some specimens more than 110 meters in height, is the so-called mountain ash *(Eucalyptus regnans)*, which grows naturally in Victoria and Tasmania. (Several authorities say that one specimen has been reliably recorded at 114 meters tall, and that others of even greater heights have been reported; if so, these would easily exceed the tallest-known North American coast redwood to become the tallest trees on earth.) Many other eucalypts exceed 46 meters in height and 20 meters in circumference. Most individual species occur only in small sectors of the country. Only one, the red river gum *(E. camaldulensis)*, has extensive distribution and is found in about half the continent. Occasionally other trees mix with the dominant eucalypts; these include box *(Tristania)*, *Acacia*, *Casuarina*, and Cypress pine *(Callitris)*, a relative of the South American *Araucaria* and *Nothofagus*.

Within these woods and across the continent live a multitude of birds, some 700 species throughout Australia. The largest are the emus *(Dromaius novaehollandiae)*, 2 meters tall, flightless three-toed inhabitants of the bush steppes, and the closely related cassowaries (*Casuarius* spp.) of the dense rain forests. Cassowaries, which weigh about 85 kilograms, subsist on fallen fruit and small animals. Even though huge, however, these nondescript birds are seldom easy to observe; Australia is more celebrated for the many conspicuously colored birds in its woodlands.

We scarcely expected to find parrots at an altitude of 3,000 meters in the Snowy Range of southeastern Australia, but while among the snow gum *(E. pauciflora)* forests of Kosciusko National Park we often glimpsed flocks of

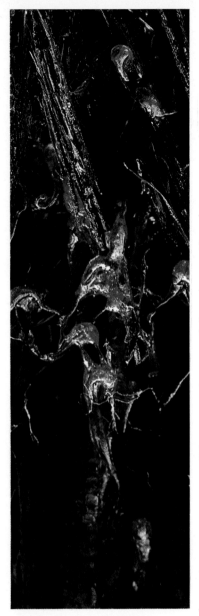

98–99. *Rainbow lorikeets*
(Trichoglossus haematodus),
*shown at the Corrumbin
sanctuary in Queensland, are
swift-flying arboreal parrots that
frequent coastal lowlands in
search of flowering trees and
shrubs. With their specialized
brush-like tongues they feed on
nectar. Scores of other brightly
colored parrots inhabit Australia.*

crimson rosellas *(Platycercus elegans)* flying across snow-covered meadows. At lower elevations and in warmer environments, Australian parrots have evolved into extraordinary forms endowed with brilliant colors. One group, the lorikeets, developed specialized brush-like tongues for feeding on nectar and became largely arboreal, especially among the abundant flowering trees of coastal lowlands in eastern Australia, New Guinea, and Indonesia. One of the most striking, the rainbow lorikeet *(Trichoglossus haemotodus)* has a blue head, bright green tail and upper parts, coral bill, green collar, and orange-red breast.

The cockatoos, a second group that spends more time on the ground than lorikeets do, have crests and short square tails. The glossy cockatoo *(Calyptorhynchus lathami)* feeds almost exclusively on the seeds of *Casuarina* trees. A third group includes other parrots such as the spectacular eclectus *(Eclectus roratus)* and king *(Alisterus scapularis;* with males and females of different brilliant colors) and bluebonnet *(Psephotus haematogaster)*. Australia possesses other remarkably patterned birds, such as kingfishers, dollar birds *(Eurystomas orientalis)*, rainbow birds *(Merops ornatus)*, finches, chats, wrens, and pittas, which combine variously in eucalyptus groves and notable flowering trees to make these forests among the world's most colorful.

The Lyrebird in Sherbrooke Forest
To seek out the lyrebird we descended into a ravine in Sherbrooke Forest, outside Melbourne. It seemed as if we were plunging into a cave: darkness closed in around us; layer after layer of giant tree ferns, their leaves almost locked together to shut out the sun, rose above us. Whatever impressions we may have had of Australia as a land of bright, hot desert soon vanished in the gloom of this wooded gully in Victoria. We plowed through undergrowth and stumbled over steeply tilted logs hidden in the darkness. For a moment all was silence; not a creature could be seen or heard. Then a piercing shriek filled the vale, then another, and another in rapid succession, and we knew we had found what we came to see.

First the call of a kookaburra rang out; then of a black cockatoo, next a currawong, whipbird, thrush fantail, robin, honeyeater, magpie, and bowerbird, all in rapid succession, all mixed with what seemed like coughs, wheezes, sneezes, and whistles. Were there actually a dozen birds in the ravine? No, just one. The world's greatest mimic: a lyrebird. Had we not been accompanied by Dr. L. H. Smith, an authority on the superb lyrebird *(Menura superba)*, we would never have believed that a single bird could utter so many sounds.

When we finally caught sight of it, the versatile bird seemed little more than a nondescript gray male nearly a meter in length, standing on a mound in a small forest opening it had cleared by itself. Here the polygamous male performs its mating dances, accompanied by a rich outpouring of high-decibel melody. As Smith, who has virtually lived with the birds, describes this spectacle: "Suddenly there arises a whirring snapping sound—it is

With more than 500 recorded species and many hybrids, eucalyptus is the dominant group of trees of Australia, growing in every environment except the most arid interior deserts and the deepest tropical rain forests. Their flowers furnish most of the honey of Australia.
Row 1, left. Eucalyptus exima, *New South Wales.* Center. *Square-fruited mallee* (Eucalyptus tetraptera), *New South Wales.* Right. *Swamp bloodwood* (Eucalyptus ptychocarpa), *Northern Territory.* Row 2, left. *White box* (Eucalyptus albens), *New South Wales.* Center. *Bell-fruited mallee* (Eucalyptus preissiana), *the Stirling Range of Western Australia.* Right. *Darwin woollybutt* (Eucalyptus miniata).

Left. *Snow gum* (Eucalyptus niphophela), *in Kosciusko National Park, New South Wales, grows at higher elevations than any other Australian tree and thus is often dwarfed.*
100. *Yellow stringybark* (Eucalyptus muelleriana), *growing here at Wingan Inlet, Victoria, has fibrous bark that peels off in long, wide strips.*

*One of the most spectacular avian courtship displays is that of the male lyrebird (*Menura superba*), found with his drab female partner in the mountain forests and fern gullies of southeastern Australia.*
Above. *In Sherbrooke Forest in New South Wales, the male raises its intricate and delicate tail feathers over its head and displays them directly in front.*
Right. *Before opening its tail, the male sings on a display mound and mimics a remarkable variety of other birds.*

Most forest marsupials are shy
and lead hidden lives among the
trees; many are also nocturnal, so
that even nearby human dwellers
seldom see them.
Above. *The greater glider*
(Schoinobates volans), *shown in
Queensland, is a nocturnal
marsupial phalanger that can
soar for as much as 110 meters
from tree to tree. Measuring about
90 centimeters long, it is the
largest flying marsupial in
Australia, as its name suggests.*
Right. *The tree kangaroos*
(Dendrolagus) *have evolved to live
in the dense rain-forest canopy.
Sharp claws, foot pads, and
elongated forelimbs enable them
to leap agilely from branch to
branch.*

the prelude to the display. In another instant the bird's long tail is thrown forward over his head, concealing his body beneath a canopy of shimmering silvery filaments. The sight is breathtaking; the transformation from the plain brownish-black bird to this spectacle of swaying thread-like plumage is almost incredible. You see no body; you merely hear a voice clothed, as it were, in a silver raiment." Smith claims that one bird he watched imitated 27 different sounds, including other birds, koalas *(Phascolarctos cinereus)*, and even railroad trains. Little wonder Australian naturalists refer to the matchless lyrebird as the "prince of mockingbirds."

A Multitude of Marsupials
Australian forests nourish numerous pouch-bearing mammals called marsupials, which give birth to underdeveloped young, then complete the rearing process in a pouch, or marsupium.

The best-known marsupials are the kangaroos, of which there are 51 species and 93 subspecies, varying from rat-size to man-size. Most familiar are the great gray kangaroos *(Macropus giganteus)*, red kangaroos *(M. rufus)*, and wallaroos *(M. robustus)*. The way they move and their herbivorous appetites adapt them best to grassy and shrub regions or open woodlands; tree kangaroos, however, inhabit forests of northwestern Queensland and New Guinea. They seem peculiarly awkward, even though their arms and hands are more strongly developed than those of other kangaroos. Their slow movements suggest that predators have never chased them from branch to branch; hence they did not need to develop agility as an escape mechanism.

The marsupials also include smaller creatures such as pouched "mice" *(Phascogale spp.)*, gliders, and cuscus and larger ones like wombats *(Phascolomis hirsutus)*, the latter weighing up to 35 kilograms. Wombats look like huge moles, with long claws that aid in digging tunnels, which sometimes exceed 12 meters in length and provide safe places to locate nests and living quarters. Although they like to sunbathe occasionally, wombats come out mostly at night to feed, cropping grass, roots, and fungi with sharp incisor teeth that grow continually.

Many other marsupials become active only at night and are rarely seen by human observers. Despite their nocturnal habits, some of these have been assiduously hunted and brought to the brink of extinction.

Gliders in the Night
In the inexorable progress of evolution, life forms have continuously improved, or at least varied, in accordance with environmental demands. Animals of dense forests, such as gibbons and monkeys, have little difficulty in sustaining rapid travel through trees. But in Australian forests the trees are not always close together; and if an animal had to descend to the ground in order to reach the next grove, it would face the danger of predators. For certain marsupials such difficulties have, over the years, been resolved by an ingenious adaptive device: membranes along the sides of the body. When flared, these folds of skin

Top. *The squirrel glider* (Petaurus norfolcensis) *fills an ecological niche in Australia occupied by squirrels on other continents. Gliding from tree to tree without ever descending to the ground permits these otherwise vulnerable marsupials to avoid predators.*
Above. *The brush-tailed possum* (Trichosurus vulpecula), *a common nocturnal phalanger, has gradually become adapted to feeding in suburban gardens as well as on tender foliage in wild forests.*

107. *The sedentary koala (Phascolarctos cinereus), shown here in Queensland, eats nearly a kilo of leaves per day, selecting its supply from only a dozen kinds of eucalyptus, and rarely drinks water. The female bears a single young, which leaves the pouch when about six months old.*

form a "flying carpet" that enables the animal to soar from limb to limb like flying squirrels of other continents. Pygmy gliders *(Acrobates pygmaeus)* make only a short glide, perhaps for one meter. But greater gliders *(Schoinobates volans)* may soar for more than a hundred meters, with their fluffy tails serving as rudders. Sometimes their glides are accompanied by loud gurgling cries, especially during the mating season. The reasons for these cries, however, are not fully understood; they may be a method of fixing the location of an individual in a group. One remarkable aspect of these gliding activities is that they occur mostly at night. Large, light-gathering eyes help these adept animals avoid collisions during their nocturnal flights. Another valuable adaptation has arisen from the considerable speciation among the 44 species of Australian phalangers, both gliding and nongliding, that inhabit the treetops. Their food requirements have become more or less distinct: pygmy gliders seek insects and flowers; medium-sized gliders live on plants, mice, and insects; the greater glider is almost entirely vegetarian.

Koalas: Captives of the Eucalyptus

Koalas nearly disappeared after great numbers were killed for their fur in the first third of the twentieth century. This slow-moving marsupial has a difficult time even under natural conditions, and its life is not entirely idyllic. Few mammals are so dependent upon trees, and of the 500 or so species of eucalypt in Australia, these creatures will consume the leaves of scarcely a dozen. Getting moisture from the leaves, they have no need to drink—which is why the aborigines called them *koolahs*, meaning "no drink." They have no weapons, no natural enemies, no need to fight, and hence no need to move very fast. Thus vulnerable, they were caught with ease by human beings and their hides shipped to foreign countries. But pneumonia and other diseases, together with ticks, internal parasites, digestive upsets, drought, and introduced foxes, already had caused koalas enough problems without human interference. Moreover, being slow-moving and confined to trees, they got caught in forest fires, and no one can tell how many thousands or millions have been burned to death.

By 1908 the State of Victoria had banned the killing of koalas and taken measures to preserve what was left of their natural habitat in Wilson's Promontory National Park. As a result of a koala management program, animals raised in special sanctuaries were returned to large areas from which they had been eliminated. Now these animals are protected in every Australian state and have become almost a national symbol. The extinction of these gentle marsupials has thus been averted—these uncommon creatures that hardly, if ever, come down from the trees.

New Zealand: Kiwi Country

New Zealand lacks Australia's great variety of mammals but has perhaps a more spectacular landscape, as is found on the North Island: erupting volcanoes, snow-capped peaks, hot springs and lakes, steam vents, moist grassy meadows, mountain streams and dense forests.

New Zealand's natural forests are like temperate rain forests. Giant tree ferns tower overhead up to 8 meters. Bright-flowered rata vines *(Metrosideros robusta* and *M. lucida)* curl among the undergrowth and climb into the forest canopy, imparting flashes of red or white with their delicate blossoms. Another vine, the bush clematis, or puawhanganga, climbs by means of strong tendrils. Yellow blooms of kowhai bushes *(Sophora microphylla)* illumine the shadows. Some native forests possess red and silver beeches *(Nothofagus fusca* and *N. menziesii)* and other species, while North American lodgepole pine *(Pinus contorta)* has invaded the lower grassland slopes. Elsewhere are remnants of coastal rain forests and the carnivorous giant land snail *(Paryphanta hochstetteri)*. From high points one can view mountain ranges and ravines covered with dense woods. The famed Mount Cook National Park, principally alpine and glacial, has forested ridges and rich moist meadows. It is adjoined by Westland National Park on the other side of the high mountains, where steep, rich rain forests sometimes get drenched with 250 millimeters of precipitation within 24 hours. Here we find one of the most spectacular flowering trees, the southern rata *(Metrosideros umbellata)*. On the southwest tip of the South Island silver beech makes up most of the forest, but the moist climate nourishes a rich vegetation that sometimes clings to steep cliffs of fjords.

Forest parrots, including keas *(Nestor notabilis)* and kakas *(N. meridionalis)*, inhabit the woods, along with rosellas and cockatoos from Australia. Other exotic birds have plagued New Zealand: California quail *(Lophortyx californica)*, English rook *(Corvus frugilegus)*, myna, and goldfinch.

The favorite local bird is the kiwi *(Apteryx australis)*. Indeed, New Zealanders often refer to themselves colloquially as "Kiwis." This long-billed, tailless, flightless bird, a maximum of 84 centimeters long and 35 centimeters high, inhabits wooded regions up into the mountains. After spending the day at rest in caves surrounded by dense vegetation, it comes out at night to hunt for fallen fruits, insects, larvae, and worms, often plunging its beak into the soft ground in its search.

New Zealand's Flightless Birds
The kiwi, the only one of New Zealand's birds that cannot fly, prompts us to ask: Why can't they? Birds ought to fly, one thinks, and so we wonder what evolutionary process brought these birds to this anomalous state. One answer is that they haven't needed to; in the absence of predatory mammals on these islands, the birds took over niches occupied on other continents by ground-dwelling life forms. Never having had to rely on flight to escape an enemy, their ability to fly gradually diminished or was lost altogether.

The kakapo *(Strigops habroptilus)*—kakapo means "green parrot" in the language of the Maori—glides but cannot fly. Indeed, this proves to be a practical arrangement, since the bird comes out at night to climb trees in search of food or nectar and then glides nearly 30 meters back down to the ground if necessary. Rediscovered only recently, the

kakapo has been crowded by habitat destruction and by domestic cats and pigs into its last refuge, the region encompassed by Fjordland National Park, at the southwest end of the South Island.

A flightless rail, the weka *(Gallirallus australis)* is a curious, omnivorous bird which runs and swims rapidly and is active at dusk in dry scrub regions or the forest edge. Another flightless rail, long the object of search and for a time feared to be extinct, the takahe *(Notornis mantelli)*, was rediscovered in 1948 and today inhabits the Murchison and Kepler ranges. With its showy iridescent blue, green, and scarlet feathers, this large gallinule has become a symbol of New Zealand's bird life on the brink of survival.

Islands of the South Pacific

They may seem much alike when approached from the sea. Inevitably a fringe of coconut palms *(Cocos nucifera)*, and perhaps pandanus, lines the shore. But when one flies over them, as we did, from New Zealand to Tahiti, the islands of the South Pacific reveal distinguishing characteristics.

To be sure, there is much mixed tropical rain forest that extends from New Guinea to the Melanesian archipelagos of the Solomon Islands, New Hebrides, and Fiji, and then eastward to Samoa, Tonga, and Polynesia. The climate grows drier as one proceeds to the east, so much so that Easter Island, some 3,000 kilometers west of the Chilean coast, is an outpost of grassland.

Typical rain forests grow in wet and sheltered places, with multiple layers of woody species; levels of clinging epiphytes, saprophytes, and vines; on down to fungi and bacteria. Termites recycle the wood. Lush growths of plants strangle other plants. Bats and insectivorous birds sweep through the upper parts of these cathedral-like chambers. These tropics shelter venomous species deadly to human beings, foremost among them being the death adder *(Acanthophis antarcticus)* and the taipan *(Oxyuranus scutellatus)*. More often seen, however, are the harmless reptiles and geckos, small nocturnal lizards that specialize in capturing insects.

Mammals are rare or nonexistent on many islands of the South Pacific (except, of course, for those species that have been introduced from other lands by man). An interesting native, which a newcomer might mistake for a monkey or a lemur, is a marsupial, the cuscus *(Phalanger maculatus)*. It spends most of its time in trees, with its long scaly tail curled around a branch for support. When it does move, the pace is apt to be slow and methodical, like that of the koala. It searches for fruits, leaves, insects, and perhaps even birds and birds' eggs, but in its habitat never has to worry about such large mammalian predators as leopards and jaguars.

Wildlife on these isles of paradise, so distant from large landmasses, will probably always remain fairly scant— only 16 species of birds in all on Tahiti, for example, and mammals even less common. But perhaps the great biomass of the sea compensates. And if a forest has less than abundant wildlife, it still serves well and vitally as storage for rain, shelter for organisms, and a milieu for fertile soil that is so necessary in sustaining human life.

108 top. *The double-wattled cassowary* (Casuarius casuarius), *a flightless bird, 1.5 meters tall, found only in dense rain forests along the northeast coast of Australia and in New Guinea, feeds largely on berries and palm seeds.*
Center. *The North Island kiwi* (Apteryx australis), *a shy nocturnal forest dweller, typifies the flightless birds that have evolved in the relative isolation of New Zealand.*
Bottom. *The crowned pigeon* (Goura cristata) *of western New Guinea and neighboring islands, the world's largest pigeon, forages on the ground for berries, fruits, and seeds.*

109–111. *New Guinea is the home of some 40 species of birds whose ornate plumage and bizarre displays are unrivaled in the avian world. Raggiana bird-of-paradise* (Paradisaea raggiana), *like others of its family, performs a flamboyant display to attract the plainer female.*

Major Types of Forest Vegetation

	Montane
	Deciduous
	Tropical Rain
	Mediterranean Scrub

Africa's Living Forests

They once called it the "Dark Continent," land of shadowed mystery and ancient menace, where "geographers, in Afrik maps, with savage pictures fill their gaps." But in reality, it blazes with light and color and almost seems to vibrate with the exuberance of life. All this is not the contradiction it seems, for there is not merely one Africa but a hundred, from islands to coasts to deserts, from mountain glaciers to winding tropical rivers, and from seemingly endless plains to dense, nearly impenetrable forests. Everywhere the African continent is a collection of contrasts.

Tropical rain forests grow on a relatively small percentage of African land, mostly in the Zaïre River basin and along the Gulf of Guinea. North, south, and east of this area the woods stretch out to become thinly forested savannas, woodland and shrub associations, or sparsely vegetated arid regions. Montane forests, much less extensive than on other continents, occur principally in southern Africa, parts of Kenya and Tanzania, and the Ethiopian highlands. A comparatively narrow band of Mediterranean scrub along the northern rim of the continent completes our picture of Africa's somewhat scanty forest terrain.

Africa is both soaking-wet and dry as a stone, and anyone who thinks of this continent as characterized by jungle is mistaken. Most of Africa is open, bright, and arid, the only continent with large northern *and* southern hemisphere deserts. Where trees grow in such environs, they are apt to be along rivers (as in Sudan and Senegal) or to comprise sparse desert scrub or some member of the *Euphorbia* or *Acacia* tribes, bristling with thorns. The genus *Acacia*, in the legume family, has a huge number of representatives, some 750 species worldwide. The African forms, well-adapted to regions of scant rainfall and some of them even to alkaline soil, are important sources of shade and fodder for wild and domestic animals.

Such vegetation, in addition to vast regions of dunes, dry basins, and more or less barren mountain systems, constitutes the principal aspect of the African continental landscape. Though extremely fascinating, these prevailing arid and semiarid regions contrast sharply with the wet ecosystems of central and western Africa. In the heart of this region the rain *does* pour: precipitation amounts to at least 1,200 millimeters a year, with some falling every month. These conditions lead to a predictable cycle of life, without the devastating extremes of six months' drought and six months' rain one encounters in other tropical forests, as in Central America.

A rain forest is exactly that—woods drenched by rain that falls day or night, or both, in great quantities. The roar of a typical downpour is almost deafening, and the immense flow of water, coming so often and so regularly, leaches the soil of much of its life-giving nutrients. Nevertheless, a massive vegetation grows, though perhaps not so densely as one would expect. We have found it far more difficult to walk through certain temperate coniferous forests than through some rain forests. In the tropics one is surrounded by giant trunks of trees whose leafy canopy reduces the amount of sunlight that reaches the forest floor; therefore the vegetation density tends to be greater up above than on the forest floor.

114–115. *A family of vervets*
(Cercopithecus aethiops) *huddles
in a rain forest above Victoria
Falls, Rhodesia. Vervets, the
most common monkeys in
forested parts of Africa, travel
in troops of ten or fewer, led by an
old male. They sleep in trees and
feed on fruits and leaves. When
alarmed, vervets pull leafy tree
branches around themselves for
protection.*
Chimpanzees (Pan troglodytes)
*sleep, nest, chatter, play,
reproduce, and feed within the
tropical rain-forest ecosystem,
abundantly surrounded by the
staples of their diet: blossoms,
leaves, stems, and fruit.*

Above. *A chimpanzee, most
human-like of the anthropoid
apes, is seen balancing on a tree
branch, in Zaïre. Chimps may
exhibit a variety of facial
expressions conveying such
emotions as fear, anger, or
contentment.*
117 top. *In Tanzania's Gombe
National Park a young chimpanzee nuzzles its mother.*
Bottom. *For the first two or three
years of their life, young chimps
spend most of their time with their
mothers, nursing or frolicking
with other chimps, as seen here.*

The exceptions to this general occurrence are open places,
such as along riverbanks, where there is access to sunlight.
The trees, along with the plants, mosses, ferns, vines, and
epiphytes (air plants) that grow in or on them, drip with
water and are bathed in vapors; yet sometimes the rain
forest can seem quite dry, perhaps because the abundant
water flows away so rapidly and the sunlight strikes with
such intensity afterward.

The heat, humidity, and stable environments provided
by the shadowy equatorial forests nourish trees such
as the buttressed sapele, or African mahogany *(Khaya
ivorensis)*, 50 meters tall and 6 meters around at the base.
Walking through this dense forest, we might well be
confused by the number of species and perhaps find it
difficult to determine the ecological patterns: that is, what
grows where, and how each plant relates to its neighbors,
the soil, moisture, sunlight (if any), and animal life. We
might see, typically, sapeles rising out of sight into the
leafy canopy above; ebony trees *(Diospyros ebenum)*, with
lustrous black heartwood; oil palms *(Elaeis guineensis)*,
thrusting out their feather-shaped leaves; violets
burgeoning in the rotting humus of the forest floor; white
arum lilies *(Zantedeschia aethiopica)*, luring insects to
their spathes, thereby fostering pollination; and fragrant
white-blossomed *Stephanotis* vines, a member of the
milkweed family, climbing the trunks of trees. We might
see insects or the birds called honey-guides *(Indicator
indicator)* in pursuit of nectar among the yellow flower
clusters of the bird-of-paradise plant *(Strelitzia reginae)*.

A Day in the Life of a Chimpanzee

Chimpanzees *(Pan troglodytes)* are, of course, thoroughly
at home among the limbs and vines and travel with ease
through this forest. They were made for this life: arms
longer than legs, ideal for swinging from limb to limb; hands
longer than those of gorillas *(Gorilla gorilla)*, better for
gripping limbs and vines; and feet long and slender, with an
extended big toe giving added leverage in their wanderings
through the forest canopy.

Their day begins at or near dawn with a yawn, a stretch,
and a climb down from a nest that may be 5 to 15 meters up
in a tree. A search through the undergrowth, to collect
berries, pursue small mammals, or catch insects, may be
followed by a climb to eat leaf buds or fruits. Principally
vegetarian, chimps eat just about anything available.
Because their food has such a high water content, they
scarcely have to drink. One would suppose that a few
minutes' foraging each morning among the abundant fruits
and vegetables at hand would suffice them for the day. But
a chimpanzee's feeding lasts for six to eight hours, as if it
were less a physical need than an adventure.

Later on, another chimpanzee may appear, whereupon the
two greet and begin to groom each other, probing for grass
seeds and ticks. This activity may be interrupted by distant
excited hoots suggesting that other chimps have found a
treeful of ripe fruit. Scampering and swinging, the others
get there rapidly and gorge until midday.

After that, a period of rest and grooming is combined with
a time of play for the young. But they must remain alert to

enemies such as leopards *(Panthera pardus)*, whose prowling can end their frolic. In late afternoon there may be a trip to a stream for a drink, a jump across the stream, and a scamper through the woods—all the while jabbering at each other or at the forest in general. It is a noisy community. Chimpanzees have a large vocabulary of calls that signify such emotions as fear, pleasure, or pain. There is also a complex nonverbal communication system based on gestures, touch, posture, and facial movements—many of these quite man-like.

Chimps travel in loosely knit groups, although something of a hierarchy does exist. High-ranking males use displays of ferocity—drumming on trunks, beating chests, dancing about—but fighting is rare among them. The social interactions, very complicated as may be imagined, appear to be based on a matriarchal structure, especially on the important role of mothers with young. Their notable intelligence is demonstrated, for instance, by their use of simple stick tools to extract termites from nests. There seems to be no end to the range of their activities, which lend the African forest striking vitality.

Gorillas occupy forests from the West African lowlands, beside the Gulf of Guinea, to highlands on the Virunga volcanoes in eastern Zaïre. Theirs is a lazy, peaceable life of waking, feeding, resting, and sleeping. Like chimpanzees, gorillas may move about more by means of their arms than their legs, though they do spend considerable time on the ground. Foraging on the ground permits them to vary their predominantly vegetarian diet and adapt to the availability of foods. Despite such flexibility, they need relatively large areas of wild forest for roaming.

Silent Mammals: Okapis and Antelopes

We could spend a lifetime in tropical African forests and not know all the drama they contain: the calling of hornbills, great flights of bats, fish eagles *(Cancuma vocifer)* diving for food in streams, dread mambas *(Dendraspis* spp.) and poisonous Gabon vipers *(Bitis gabonica)*, and so on. Whereas chimpanzees fill the woods with indiscriminate chatter, so that the visitor knows exactly where they are, some other mammals utter few sounds and withdraw so silently into the woods we hardly know they are there. If we stepped through these woods as quietly as an okapi *(Okapia johnstoni)*, we might encounter one of these shy relatives of the giraffes. The okapi appears to be wearing striped pants, but otherwise has a pale brown or dark gray coloration that merges with the shadows. Nearly 2 meters tall and weighing about 250 kilograms, this is no small creature; yet its camouflage and solitary, secretive habits have helped it elude man so successfully that it was not discovered until after 1900. That it feeds on the shoots of young shrubs and battles to defend its young are among the few facts known about it.

Of lesser stature than the okapi, but as rarely observed, is the bongo *(Boocercus eurycerus)*, a handsome horned ungulate with a reddish-brown coat and vivid white pattern that make it one of the most attractive antelopes. A solitary, retiring creature, it subsists on leaves, fruit, and plant shoots in dense central African forests.

Largest of the great apes, the gorilla (Gorilla gorilla) *feeds on more than a hundred different species of plants and travels continuously within its home territory, searching for food. Because the nutritional value of much plant material is fairly low, large quantities must be consumed.*

118 top. *High in a tree, an immature male mountain gorilla is seen eating.*

Bottom. *A female gorilla is caught in a menacing attitude, possibly protecting her young nearby. Generally, the gorilla is peaceable and sociable, coexisting with chimpanzees that may occupy the same range.*

Many mammals of deep African forests are difficult to study because of their shy, retiring habits and protective coloration patterns that obscure them from observation.
Above left. *The okapi (Okapia johnstoni), the giraffe's nearest living relative, lives in a small area of equatorial rain forest partly within the Ituri Forest.*

Above center. *The bongo (Boocercus eurycerus), shown here in Aberdare Forest, Kenya, is an antelope that is more widespread in Africa than the okapi. Its habitat is dense, humid forests with much undergrowth. Despite their large size, bongos can move swiftly and silently through tangled thickets. Both sexes have horns, which they use to uproot saplings to eat the roots.*
Above right. *The name duiker, meaning "diving buck," describes the action of these small forest-dwelling antelopes (Cephalophus) in darting off into deep vegetation when alarmed. Because of their shyness, dense forest habitat, and largely nocturnal feeding, duikers are seldom observed by human visitors.*

Right. *The spiral-horned sitatunga (Tragelaphus spekei), seen resting in a Cameroun rain forest, is more shaggy-haired than most other African antelopes. There are three subspecies, each of which inhabits a separate area in Central Africa. They prefer a riverine forest or swampy habitat.*

Also hard to find in the tropical rain forest are duikers (*Cephalophus* spp.), of which several species have evolved, none taller than about 45 centimeters or heavier than 65 kilograms. The habits of these, too, are little known. The red duiker *(C. natalensis)* is said to climb slanting tree trunks to browse on foliage, in the manner of goats. Though they prefer to dine on vegetation, duikers also eat termites and may occasionally kill birds. But the secrets of these and other quiet mammals of the rain forest have yet to be fully explored—impressive testimony to both the richness and the remoteness of equatorial forest ecosystems.

The Populous World of Beetles

Many insects are just as characteristic of Africa as the chimpanzee and as exotic as the baobab tree. Although insects may seem less dramatic than the huge mammals or birds, they can be fundamental to the health and welfare of other parts of African ecosystems.

If we think that differences among human beings in Africa are significant, we should recall that there is, nevertheless, only a single species of human being. Try, then, to imagine an animal evolved into well over 290,000 species worldwide. Such is the record of the Coleoptera, or beetles, which range in length from only a fraction of a millimeter up to 15 centimeters. Such specialization might seem incredible until we examine the multiple natural niches that beetles occupy: there seems to be a beetle adapted to feed on just about every kind of organic matter, dead or alive. In one sense, they "clean up" the environment; in another, they control the proliferation of other insects and prevent their overabundance. In yet another, they in turn become meals for larger animals, the insectivores. They convert wood to dust, spread pollen and seeds, trim trees, enrich soil, and perform innumerable other tasks vital to sustaining forest growth.

Men have by no means mastered the intricate details of their behavior, some of which seem uncanny. Certain African rove beetles *(Doryloxenus)*, for example, live in close association with army ants and participate in long expeditions: with its legs and mouthparts, each beetle clings to an ant for transport from place to place. Another rove beetle *(Stenus)* shoots out its sticky labium, like the flick of a chameleon's tongue, to capture prey. For hunting, the predatory tiger beetles (Cicindelidae family) have evolved huge pincers, threatening antennae, large eyes that provide acute vision for tracking prey, and an agility to scurry after and pounce on victims. The more we discover how beetles function in each tiny forest habitat, the more we understand why so many species have evolved.

Ants and Anteaters

Perhaps the finest example of plant-animal-insect adaptation is that centered around the pangolins (*Manis* spp.). Some pangolins are strongly arboreal and usually nocturnal, dependent on trees for support and on tree and ground insects (termites and ants) for sustenance. These mammals, covered with scales and protective eyelids to ward off the bites of ants and other insects, in general look like a cross between armadillos and anteaters. They have

Top to bottom: (1) *A West African scarab beetle* (Allomyrina dichotomus), *scavenging on waste and decaying matter, makes a ball of food material, rolls it along, and often buries it in the soil for use by developing larvae.* (2) Stephanorrhina guttata *is a horned beetle of the family Scarabaeidae, whose members have a wide variety of horn shapes. Horny outgrowths on many parts of its body serve for such multiple uses as burrowing and carrying food.* (3) *This long-horned beetle belongs to the family Cerambycidae, the members of which are vegetarians and include many timber-destroying insects.* (4) *A leaf beetle* (Platypria) *of Liberia illustrates the variety of forms and structures found among members of this group.*

Mantids, belonging to the order
Orthoptera, hunt by day; their
coloration serves either as
camouflage or as a lure for
potential prey.
Above. A mantid in a Zaïrean
forest strikes a defensive stance.
References to "praying" mantids
derive from positions assumed by
their prehensile front legs while
they wait to capture food.
Left. A West African mantid
(Pseudocreobotra ocellata) makes
a kill. Mantids eat other insects,
frogs, lizards, and young
birds—as well as each other.

long tails, long narrow snouts, a sticky rope-like tongue, and overlapping horny scales that act like armor when they roll up in a ball. This is not an altogether invincible stance, for they can be pried open by larger and stronger predators such as leopards.

All pangolins are insectivorous and have strong claws on their forefeet that they use to rip open nests of termites and ants. Their lack of teeth means that the 150–200 grams of termites they consume each day reach the stomach uncrushed and must be mechanically ground there. The stomach therefore has a hard laminated epithelium and tooth-like projections that do the job, possibly with some help from pebbles ingested along with the insect food. During hard times, pangolins may fast perhaps for eight weeks, but when their reserves of fat get low, they must replenish or die.

The white-bellied tree pangolin *(M. tricuspis)* is a nocturnal arboreal species of the tropical rain forest from Sierra Leone eastward to the Great Rift Valley. In some areas such as Zambia and southern Zaïre, it lives along the forest edges and out on the savannas, preferring termites to ants for its diet. The long-tailed tree pangolin *(M. tetradactyla)* remains largely in the trees but circulates in daylight hours to seek ants. The terrestrial giant pangolin *(M. gigantea)* is a night-prowling hole-digger, which uses its large claws to rip open mounds made by ground termites.

Other pangolins live in Africa and elsewhere; but from this brief introduction to a few species, we can see how precisely evolution has fitted certain animals into the forest environment and, accordingly, how disastrous any interruption along this chain of life could be. What happens to termites or ants could profoundly affect the pangolins.

And other animals as well. Since ants have been around for a hundred million years, and are such a superb source of protein, it is little wonder that a host of mammals, birds, reptiles, and insects have evolved not only to prey on them but also to depend on them for life itself. Perhaps the aspect most beneficial to pangolins, chimpanzees, and other "consumers" is the colonizing capacity of ants: all ants live in enormous communities; a fully developed colony may contain as many as a million members, supplying a convenient concentration for predators.

Though there are more than 9,500 species of ants, some have become highly specialized, and predatory animals capitalize on this by ambushing them as they march in formation or by ripping open their mounds, chambers, and passageways. Of course, they can sting, but protection against ant attacks has been a part of the evolution of their predators—pangolins being a case in point.

And if a predator doesn't like the taste of one species, it probably is not far to the colony of another species. Some ants wander, cut leaves, tend nests; some pulverize logs, snags, and stumps as a part of the forest's recycling process. Others are themselves predators, which attack insects, poison enemies, or capture slaves. Their social behavior is as fantastic as that of bees. It is incredible to witness the cooperation and precision among thousands

124. *Pangolins in Africa fill the ecological niche occupied by anteaters in South America. At Mount Nimba, Liberia, a giant pangolin* (Manis gigantea) *is seen descending a tree trunk. Its sharp claws are useful for digging into ant and termite nests.*

Termites, one of the most abundant rain-forest insects worldwide, are prolific reprocessors of dead wood and thus keep forests from becoming clogged with debris. Highly organized and efficient, they serve important forest functions, principally as soil modifiers and as food sources for other animals. Above left. *Near Lake Manyara, Tanzania, young winged males and females swarm, with the urge to fly and mate, and establish new colonies. In leaving their nest, they invite predators.*

Above center. *This cross section through a termite mound (termitary) in Monrovia, Libéria, shows the complex system of passageways and fungus gardens that feed the huge colony. Temperature control within the fungal colonies is carefully maintained by the termite workers.*
127 top. *After the colony is established, the termite queen sets about laying eggs. Her abdomen swells, and she soon dwarfs the much-smaller king, who stays with her in the well-protected, deeply hidden royal chamber, as shown by these queen and king termites* (Macrotermes bellicosus) *in Ghana.*

127 center. *Termite workers in a* Cubitermes *nest at Folgares, Angola, carry eggs to special nursery chambers. Individuals passing from eggs directly to nymphs without a pupal stage become workers, soldiers, or members of the royalty.*
Near right. *A* Cubitermes *nest in the Douala rain forest, Cameroun: The size attained by an individual nest or termitary is determined by the amount of food and water available in a given locality. There may be a thousand termitaries in less than half a hectare of rich forest.*

of ants, for example, as they serve and protect a queen of their colony.

For obvious reasons of food, warmth, and moisture, the greatest concentrations of ants are found in or near tropical rain forests. The ant fauna of such regions is characterized by a large number of species; in the basin of the Zaïre River alone may be found some 700 species of ants. Army ants of the subfamily Dorylinae and the arboreal ants *Oecophylla* and *Polyrhachis* are characteristic of the African rain forests.

Serving much the same function in the equatorial forest— but not related to ants—are the termites, sometimes mistakenly called "white ants." Most of the 2,000 species of termites are tropical. Not only do they provide a vital food base for other animals, but their huge nests also constitute shelter or vantage points. As many as 3 million termites may live in a single nest, a remarkable condition made possible by their social organization and skill at climate control inside the colony. Some termite mounds, such as those of the Macrotermitidae in Africa and Asia, are large enough to support distinct types of vegetation. Certain African termite mounds, constructed of "mud" produced from saliva and partially digested wood, may be 20 feet high. Other nest systems may be arboreal or subterranean, the latter consisting of networks of galleries radiating in all directions.

When they swarm, termites on the wing are vulnerable to attack by birds, mammals, and insects. To subsist on termites, certain mammals have become adapted in special ways: elongated muzzles, long protrusible tongues, reduced dentition, thick skins, powerful forelimbs, and strong claws. A good example is the aardvark *(Orycteropus afer)*, whose sticky tongue can be extended for up to 30 centimeters. It is even said that aardvarks can locate termites by perceiving the sound of their footfalls.

Masters of Changing Color: Chameleons
Another animal with a long sticky tongue (in some cases as long as the body itself) is the chameleon, a remarkable lizard that exists in Africa, southern Europe, and Asia in varied forms that total about 86 species. It has become adapted to arboreal life through an arrangement of its toes somewhat like that of primates, but it occupies the lower levels of tropical forests rather than the upper branches.

Chamaeleo africanus ranges across the waist of Africa. The crested chameleon *(C. cristatus)* of West Africa has a dorsal crest, emerald-green coloration, and a yellowish-red throat patch. *C. dilepsis*, which lives in tropical and southern Africa, is an aggressive species that, like its relatives, subsists principally on insects.

Most remarkable is the chameleon's ability to change color: green when among tree foliage, a tree pattern of bark color when hiding behind branches, or an "excited" coloration of black and green with white and yellow dots. While each species has a limited range of such colors, the phenomenon is nonetheless striking. In response to the motivational state of the animal and heat and light stimuli, pigment cells in the skin change color as needed.

Despite these unusual powers of camouflage, chameleons are vulnerable to predation. When the chameleon meets a boomslang *(Dispholidus typus)*, a supple snake that grows to 2 meters long, the chameleon puffs up, hisses, and displays a stippled marking. All to little avail. The boomslang hides in foliage and waits to make a quick strike.

Exotic Trees
Beyond the rain forests, Africa has some striking and unusual trees. The evergreen *Acokanthera*, a relative of oleanders and periwinkles, is so poisonous that a poacher who fell on his own poison-tipped spear died within 40 minutes. The Wakamba and Waliangulu tribes have used this poison extract, a toxic glycoside known as oubain, on their arrows for centuries. Early explorers said some Africans considered *Acokanthera* so poisonous they would not even smell its flowers. Others say that you can identify the tree by the number of insects and small rodents lying dead at its base.

The baobab *(Adansonia digitata)* looks more like a giant barrel than a tree. To meet the threat of drought, it has evolved a huge bulbous trunk—as much as 20 meters in circumference—with little woody substance but much water-storage tissue. This enables the tree to withstand seasonal extremes and live for perhaps a thousand years, though its trunk may rot into a large-hollowed center big enough to accommodate men or beasts. The tree rises perhaps 15 meters, providing a measure of security for perching birds of prey. Its size, water content, and other special adaptations enable the baobab to survive when weather conditions are grim, as during drought. So each tree, as refuge and food source, is likely to shelter something of a mini-community.

Mammals of the Night
What happens when birds of the upper, middle, and lower forest levels settle to roost and go to sleep with the coming of night? An empty ecological niche until dawn? Not at all. Small mammals take over and feed on the same types of food utilized by the birds in daylight. For all we hear of prowling leopards and other dramatic confrontations by night, certain subtle happenings in the forest better indicate the remarkable precision of evolution in enabling animals to adapt to circumstances.

In a nine-year-study of five animals in the West African equatorial forest of Gabon, Pierre Charles-Dominique observed that, though all fed on insects, fruit, and matter exuded by trees, they did so in different quantities and in different ecological niches.

The potto *(Perodicticus potto)* and two species of bush babies *(Galago demidovii* and *Euoticus elegantulus)* live up in the forest canopy, while two other species, Allen's bush baby *(G. alleni)* and the angwantibo *(Arctocebus calabarensis)*, inhabit the undergrowth. Furthermore, the three upper-level animals feed primarily on three different types of foods: one on insects, one on fruits, one on bark exudates. Of the lower two animals, one feeds on exudates and the other on insects, both near the ground. It is as if

they constitute contiguous pieces in a jigsaw puzzle, with their domains touching but not overlapping.

Charles-Dominique's observation seems almost a classic of the relationship between similar species in the same area: "Dietary specialization combined with differences in spatial distribution within the forest and with differences in body size permit these five prosimian species to coexist without engaging in interspecific competition."

That is a rather technical way of saying that each animal fills its own natural niche without disturbing the other. Of course, much the same is true of daytime species, but this unusual study gives us an insight into the very practical way that nature has managed her communities of life in the dark.

A Precarious Balance: Elephants and the Forest

The forest is crucial in the life of the African elephant (*Loxodonta africana*), for that is where the male and female mate and where the mother goes to bear her young. The forest is refuge at other times, too; in the dry season, for instance, elephants come to it for what food and water remain there. Indeed, if there are enough forests to go around, an elephant can get from woodlands all the nutriment it needs by eating fruits, bark, wood fiber, roots, tubers, and leaves. The trouble is that there are not always enough forests to supply its capacious need. Whole tracts of woodland have been destroyed by elephant herds whose numbers grew in excess of the forest's capacity to supply what they consumed.

To get enough food, many elephants have to migrate; their routes were historically well defined, and the food along them in balance with the elephant numbers. By contrast, the elephants of central and western African rain or gallery forests have long managed to supply themselves with less difficulty. Even so, the gallery forests are also disappearing. If enough forests vanish, so could the elephant, for an adult bull must consume at least 135 kilograms of vegetation every day. And elephants live about as long as human beings do.

Fortunately, the elephant adapts easily to a variety of vegetation. It also tolerates wide ranges of temperature, humidity, and altitude and can coexist with other fauna on the same terrain. Fortunately, too, African governments are attempting to do something about its often precarious situation. The 3,970,000-hectare Selous Game Reserve in Tanzania protects a varying population of wild elephants which seem to be getting along fairly well, because the reserve is large enough to accommodate them.

But in other areas the death rate of elephants has become alarming. African conservationists blame the elephant's sharp decline in these areas on poaching, encouraged by European and Asian demands for ivory. According to one report, Hong Kong alone imported over 500 tons of ivory in 1975, which represents between 25,000 and 30,000 dead elephants. Since that time Kenya has banned elephant hunting, but unless conservationists and governments continue to be vigilant, the African elephant might soon be added to the list of endangered species.

Like all forested continents, Africa has distinctive trees with striking features.
130 top. The acocanthera, or poison tree (Acokanthera), has highly toxic sap and fruit. Natives use its poison on the tips of hunting arrows.
Center. A baobab tree (Adansonia digitata) may live more than a thousand years. Its seeds, fruits, and leaves are eaten by many animals, including baboons.
Bottom. A bulbous Madagascar baobab displays the water-storing thickened trunk and short tapering branches typical of this genus. A fungus that often attacks the baobab trunk contributes to a hollowing-out process.

Above. *Typical of small nocturnal primates in African rain forests is the Senegal bush baby (Galago senegalensis). These and related species range throughout most forested and brushy regions of Africa south of the Sahara. As they leap from one tree limb to another at night, their presence can be detected by high-pitched chirping calls.*
132–133. The largest mammals of the African rain forest, elephants (Loxodonta africana), like this one in a woodland at Lake Manyara National Park, Tanzania, live in loosely organized herds where ample shade, abundant vegetation, and water are available. Many authorities believe elephants originated in the rain forest and adapted to savanna conditions only as water and forests diminished.

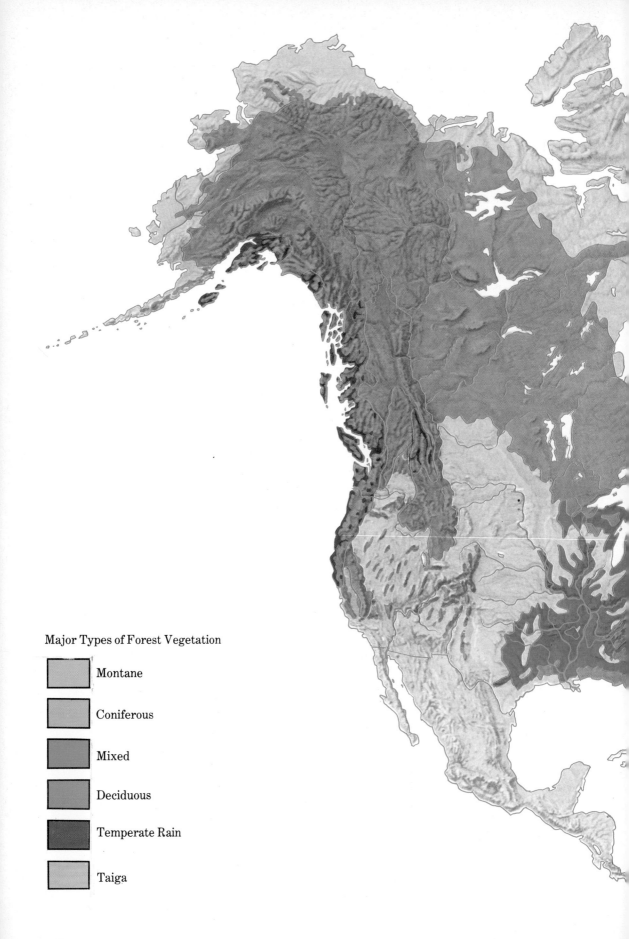

Major Types of Forest Vegetation

Montane

Coniferous

Mixed

Deciduous

Temperate Rain

Taiga

North American Wilderness

As we can see on a vegetation map of North America, the forests are widespread and diverse. A dense band of coniferous forests that include some of the world's tallest trees covers parts of the northwestern United States, Canada, and Alaska. In areas east of the Mississippi River, deciduous trees extend from Labrador westward across southern Canada and from New England down through the Appalachians. Vast swamp forests blanket much of Georgia and Florida. Other forest systems include a temperate rain forest in the northwest, cactus forests in the southwest, and mangrove forests in Florida.

Redwoods to Water Oaks:
The Variety of American Forests

The densest growth of coniferous forest in the world occurs in the northwestern United States, mainly because about a half meter of the total annual precipitation there falls in the summer. This results in an ecosystem endowed with an abundance of plants and animals, for the most part within 50 kilometers of the Pacific coast. A redwood *(Sequoia sempervirens)*, the world's tallest conifer and possibly the tallest tree, growing at Redwood Creek, north of Eureka, California, measures 111 meters in height, 4 meters in base diameter, and 13 meters in circumference. The giant sequoias *(Sequoiadendron giganteum)* of the Sierra Nevada, in central California, are even more massive and older than the coast redwoods, but the ecosystem of which they are a part is not so luxuriant.

While redwoods encompass some of the tallest, largest trees on earth, North American forests have many other fascinating attributes to distinguish them. From the imposing verdant forests of the Sierra Nevada in California, one descends to the desert regions of the southwest and there finds immense groves of succulent saguaro cactus *(Cereus giganteus)*, which may grow to 15 meters in height. At higher elevations in the semiarid regions of North America, particularly on the Colorado Plateau, are found so-called pygmy forests of piñon *(Pinus edulis)* and juniper *(Juniperus communis)* and an extensive forest of yellow pine *(P. ponderosa)* occupying these uplands.

In the Rockies as well as the Sierra Nevadas, we encounter broad expanses of lodgepole pine *(P. contorta)* and whitebark pine *(P. albicaulis)*. In Wheeler Peak, Nevada, a bristlecone pine *(P. aristata)* is estimated to be 4,900 years old, the oldest-known living tree.

Across broad grasslands toward the east, we enter the Mississippi River basin and its typical deciduous forest and meadow systems. The richness of these once-extensive woods is illustrated by the Big Thicket of eastern Texas, which at one time covered 1.5 million hectares, but only 2 percent of which remains. Although the density of woods is not always so great as the term "thicket" suggests, the number of species present and the very location of this forest are significant. Comprising a kind of vegetative "crossroads," the Big Thicket lies at or near the western limit of eastern deciduous trees, the southern limit of northern species, the eastern limit of western species, and the northern limit of certain subtropical life forms.

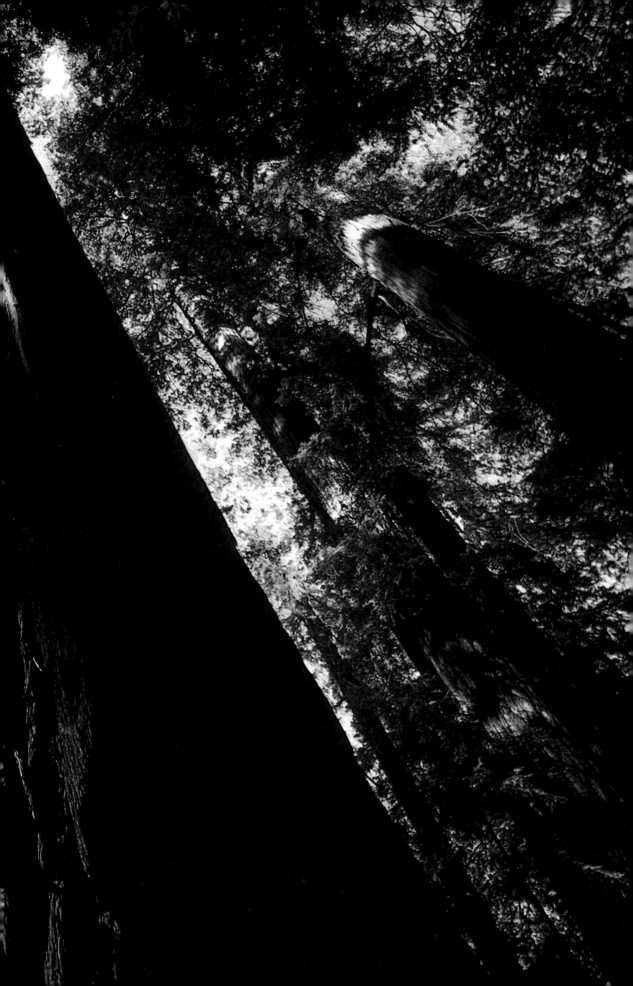

Still, the Big Thicket is essentially a southern forest, containing oaks, magnolias, palmettos, and bald cypresses *(Taxodium distichum)*, as well as American beech *(Fagus grandifolia)*.

Along the lower Mississippi River, where floods are a regular part of the natural order of life, a dense bottomland forest was thoroughly cut over, although some patriarchs up to 37 meters tall still stand. Along old river channels and beside the coastal waterways called bayous, massive bald cypress rise out of mirror-like brown waters. The enormous productivity of these survivors is suggested by one old water oak *(Quercus nigra)* that was found to produce more than 28,000 acorns in a single year. Such a collection of hardy deciduous trees constitutes what might be called an aerial infrastructure, not only for birds but also for such other animals as raccoons and opossums.

Prowlers of Woods and Streams: Raccoons and Opossums

It is not a "washing bear," as the German name *Waschbär* implies, because it is not a bear—it is actually not much larger than the domestic cat and does not always wash its food. But the raccoon *(Procyon lotor)* spends so much time near streams, wetting and fondling its foods, turning morsels over and over, then chewing them carefully before swallowing, that one can easily get the impression that this little woodland omnivore is a very meticulous creature. In fact, its sense of feeling is so well developed that it spends much time examining everything it touches.

Streambanks at night are its proper habitat; behind the rod-and-cone layer of its eyes the raccoon has a reflective membrane called a tapetum lucidum that enhances its nocturnal vision and allows it to prowl in reasonable safety. Still, a raccoon must stay alert, for it is tempting food for owls, foxes, and other predators.

With its striking mask-like markings across the face and its thick coat, which was once very popular for fur coats, it has become one of the most familiar wild animals in America. Bold and persistent, raccoons range from Panama to southeastern Canada and are moving slowly northward.

When prowling for food, they will take just about anything their dextrous fingers tell them is edible. Mostly, they explore the banks of streams and small pools in search of crayfish, minnows, dead fish, small rodents, and the eggs of ducks and swans. But they do not hesitate to invade neighboring forest uplands and fields, where they feed on wild berries, nuts, insects, and farmers' grain crops.

Dainty feeding turns into outright gluttony in autumn as the raccoon stores up a layer of fat to help see it through winter, when the streams may be frozen and natural foods scarcer than usual. That is a season to find a hollow tree, stump, or cave in which to retreat into a sort of lethargy for perhaps weeks at a time. Although this is not true hibernation, it does help the animal avoid some of winter's rigors.

With the first signs of spring, it is time to mate, and when the young arrive—usually four or five—the mother and offspring proceed on their lively, noisy explorations.

With one important exception, the raccoon (**Procyon lotor**) *uses trees for everything: sleeping, resting, refuge, and denning. But when eating time comes, trees offer only supplements to its diet; the real action takes place at the edges of ponds and streams.*
Above. *This adult raccoon, in a tree in Michigan, may be looking for ripe fruits or nuts to vary its diet.*
Left. *Young raccoons, snug in their tree den in Louisiana, are nursed by their mother until about 14 weeks old. Raccoon young also spend much time learning climbing techniques.*

Above. *Opossums* (Didelphis virginiana), *the only North American marsupials, inhabit farm areas and woodlands along streams. When the young leave their mother's pouch, they may frequently be seen riding on her back as she feeds at night.*
Right. *One of the largest remnants of more or less natural hardwood forest in eastern North America thrives within Great Smoky Mountains National Park. Some 1,300 kinds of flowering plants grow in this dense, humid expanse from which rise smoke-like mists. In late May and June, thickets of* Rhododendron catawbiensis *bloom on the rich green slopes.*

As long as parasites leave them alone, and fire or chain saws do not devastate their domain, raccoons remain one of North America's most curious and affable mammals. About the size of a raccoon, with long coarse fur, the opossum *(Didelphis virginiana)* is the only North American marsupial (others occur in Central America). It prowls by night, then sleeps by day in any shelter it can find—hollow tree, abandoned squirrel nest, woodpile, or even the basements and attics of human dwellings. The prehensile tail has numerous uses beyond helping the opossum move safely among tree branches; for example, the animal carries nesting material rolled up in its tail. Its poor eyesight is offset by good hearing, a keen sense of smell, and sensitive facial vibrissae, fibers that enable the opossum to avoid colliding with solid objects in the dark. When danger threatens, the animal can also protect itself by inducing a state of catalepsy, "playing dead"—a behavior that fools even pursuing dogs.

The female opossum carries its young in a prominent fur-lined pouch on its belly. From conception to birth takes less than 13 days; this gestation period is so short, and the young are so poorly developed, that all they can do is move a few centimeters through their mother's fur and into her pouch, find a nipple and attach themselves, then suckle for another 60 days. The newborn opossums have a set of claws that aids them in this trip into the pouch, but once they are inside and firmly affixed to the nipple, their embryonic claws drop off.

The opossum has a good survival quotient, since this birth process normally occurs twice a year and a litter consists of eight to ten young. After nearly three months in the pouch, the young are weaned and enter a period of riding on their mother's back while learning how to feed for themselves.

With 50 teeth, opossums can handle a variety of foods, from snails to seeds to beetles, and even small mammals. Feeding must continue all winter, and this necessity may force the family to resort to carrion. If the weather gets too cold, they can enter a prolonged sleep, but this is not true hibernation. Finding a hollow tree in which to den helps keep them away from prowling hungry dogs or foxes.

The Wide-Ranging Black Bears

The forests of the Great Smoky Mountains in eastern Tennessee and western North Carolina rank among the most ancient in North America simply because they were never disturbed by the advance of glaciers. In these areas humidity is high, moisture is abundant, and soils are rich; the growing season is long, and the temperature generally mild. Thus these mountains became a refuge when glaciers gripped other parts of the continental landmass. Ancient migrations across land connections that no longer exist also helped enrich the vegetation of the Great Smokies. When the glacial ice to the north finally retreated, plant life radiated out from the Great Smokies to occupy other parts of North America. Today, the woodlands of the Smokies contain rich forests: deciduous trees on the lower slopes, coniferous in the highlands.

One of the experiences of a lifetime for many visitors to the

141–143. *Black bears* (Ursus americanus), *ever-curious and ever-hungry, roam the North American woods in search of anything edible. Cubs scamper up trees and take time out to play, but food is the abiding object of life. Poor eyesight doesn't seem to matter; sharp ears and a sensitive nose lead them surely to such delicacies as deposits of wild honey.*

Smokies is a glimpse of a black bear (*Ursus americanus*).
These ponderous animals, some weighing more than
200 kilograms, inhabit an extensive range, from desert
mountains to arctic Canada and from sea level to mountain
coniferous forests. They are the smallest, most numerous,
and most widely dispersed of North American bears.
Pursuit of food is their each day's goal: collecting berries,
scooping honey from hives, digging up roots or rodents,
consuming acorns and nuts, climbing trees for seeds,
hunting for carrion. Much of this activity occurs at dusk,
at night, or toward dawn. In due course the fattened
bears retreat for a winter's sleep. This is not a true,
uninterrupted hibernation, for the bears may wake and
wander at times if a winter is unseasonably mild
or if they live in warmer latitudes. Their temperature,
breathing, and heartbeat remain more or less normal
throughout.

The black bear has an unusual breeding cycle that
undoubtedly enhances its survival factor. Mating takes
place in June or July, but the fertilized egg or ovum does
not attach itself to the uterus wall until the following
November. This delay in development of the fetus means
the young will be born in January or February, while the
mother is sequestered for her winter sleep.

Newborn cubs, usually a pair, measure scarcely 20 centimeters in length and weigh about 500 grams—that is,
approximately 1/500th of their mother's weight, the
smallest ratio of mother to young in any mammal except
marsupials. Small, blind, and weak, the young must
nevertheless find their way to her teats for food, since the
mother bear may be of no help if the birth process occurs
without waking her. As the summer passes, life for the
young becomes idyllic; black bear cubs demonstrate some
of the most active play behavior of any mammals, including
much vigorous scampering up and down trees.

When full-grown, the black bear is not always black; some
individuals display variations of tan, brown, and cream.
This will not be surprising to anyone who knows of the wide
variations in color among their relatives, the polar bears,
grizzlies, and Alaskan brown bears.

Adult black bears may look and act sluggish but can move
like lightning when an enemy—a mountain lion, grizzly,
porcupine, or man—endangers their cubs. They are
usually content to live and let live, but if provoked, they
may inflict severe injury and even cause death.

The Wood Duck

Continuing south into Georgia and Florida, we enter
extensive swamp forests, the most impressive of which is
in Okefenokee Swamp. Many tree roots have become
embedded not in solid ground but in floating peat, which
permits whole sections of forest to be joggled by
footsteps—an unsettling experience that explains their
Choctaw Indian name meaning "land of the trembling
earth," from which "Okefenokee" is derived.

One colorful and conspicuous bird of southern forests
almost met the fate of the extinct, or nearly extinct, ivory-
billed woodpecker (*Campephilus principalis*). The wood
duck (*Aix sponsa*) suffered so severely from hunting and

habitat destruction that it had approached extinction until laws prohibited these threatening practices.

Wood ducks nest in the hollows of tree trunks. Why this happens is still open to conjecture. The bird spends much time in trees and, with its body markings, is usually better camouflaged in woods than on water. Tree hollows furnish fairly safe refuges, provided raccoons cannot enter, and it thus seems logical that the wood duck, occupying an upper forest ecological niche, would nest there.

Laying and hatching wood duck eggs in a hole in a tree means that the dozen or so ducklings must eventually climb up out of their nest, which can be as much as 2 meters deep. Well-equipped with sharp, hooked claws, they are prepared to do this; but at the age they leave the nest they are still ill-equipped to fly. The mother calls; the young clamber to the nest opening and jump out, fluttering helplessly through the air to strike the ground, sometimes 20 meters below, with a relatively harmless thump. The mother collects them and leads them to water. But by the time they reach adulthood, wood ducks have developed into skillful fliers, capable of winging at high speed through dense woods, cane thickets, and tangles of vines.

Palmettos and Armadillos

The gentle climate of the southeastern United States supports a considerable growth of pine trees. On limestone ridges in southern Florida, the Caribbean—or slash—pine *(Pinus elliottii)* grows with an understory of saw palmetto *(Serenoa repens)*. These pines thrive best in elevated sites and, though they must have some moisture, cannot tolerate much flooding. They are resistant to fire and lightning because their many-layered bark protects delicate inner tissue from heat damage. Saw palmettos, with deep storage roots protected in rocky pockets, sprout again even when fire burns off most of the plant. In one sense, therefore, slash pines and saw palmettos constitute a fireproof forest. Finally around the southern coasts of Florida, and even marching into the sea, are forests of mangrove *(Rhizophora mangle)*, with its distinctive stilt-legged roots growing out of salt water.

From time to time, one may see scurrying among the palmettos a creature that looks one-third mammal, one-third reptile, and one-third something out of delirium tremens. The armadillo *(Dasypus novemcinctus)* is all mammal, however, and marvelously fitted as are few other mammals with an armor of small horny plates arranged in bands that slide against each other. Usually the dense vegetation of its habitat gives it ample environmental protection, except against cold, and since the animal has no warm coat like its mammalian brethren, its range is limited to the southern tier of states from Texas to Florida. During its nocturnal ramblings the armadillo seeks insects (mainly beetles), earthworms, spiders, snails, vegetation, and carrion on which to feed. Human observers often hear the armadillo before seeing it, owing to its grunting noises and the rustle of bushes through which it passes.

Abundant Ungulate: The White-tailed Deer

The armadillo's domain is shared with some other animal

146. Ichneumon wasps (Ichneumonidae) seek sites on dead or dying coniferous trees where wood wasps have drilled holes and laid their eggs. If an ichneumon drills into the spot occupied by a wood-borer grub, she will extrude her own eggs into it because it will provide a source of food for her own developing larvae.

No deer is more widely distributed in North America than the white-tailed (Odocoileus virginianus), growing somewhat larger in northern forests than in southern. Above. By mid-September white-tailed bucks, like this one in Canada, begin to rub their itching antlers against small trees and saplings. Peeling away the dead covering tissue, or "velvet," may take as little as a single night or as long as three weeks, but the deer is then ready for the battles of rutting.
148–149. White-tailed bucks, flashing their alarm signals—a flaring of white hairs on their rump—bound off through the Louisiana woods in search of cover. Their silent signal, readily noted by other herd members, serves all as a warning of approaching danger.

inhabitants, notably the white-tailed deer, the most abundant wild ungulate in North America. More than 5 million whitetails live today in forests and meadows from southern Canada throughout the eastern three-fourths of the United States and nearly all of Central America. They normally require 1 to 5 kilograms of browse daily, but sometimes there are not enough acorns, beechnuts, buds, shoots, and other vegetation to go around, especially since most whitetails spend their lives in an area of less than 3 square kilometers. In northerly regions, when they "yard up" in winter (i.e., gather in large groups in sheltered places such as swamps or thickets, where snow is less deep), they may soon exhaust the available food and, if trapped by deep snow, be unable to leave. Starvation results—an age-old natural process for them—and efforts by human beings to feed deer artificially only compound their population problem.

Ordinarily, however, deer feed in early morning and late afternoon; bounding away if danger threatens, they show their familiar trademark as they leap—the raised white tail flying like a flag. Their reddish summer coat, contrasting with their surroundings, is easy to spot in open meadows. But with the coming of autumn, the reddish coat yields to gray and longer, denser, more insulating fur. Altogether, some 39 subspecies of these deer have been recognized. The diminutive key deer (weighing only up to 23 kilograms, compared with a maximum of 125 kilograms for the more common variety) inhabits islands bordering Florida Bay. Though once nearly extinct, it has been carefully protected since 1957 and appears to be making a comeback.

Copperheads to Woodpeckers: The Forested Eastern Plain

In a vast arc stretching some 3,000 kilometers from New England to Texas lie the rolling Piedmont hill country and broad coastal plain of eastern North America. Much of this region is sandy and covered with oak, pine, and beech in relatively open forests dotted with abundant grassy meadows. Even though virtually the entire region has been thoroughly settled, wildlife has made a remarkable comeback here. Much land remains in a more or less natural condition because of the numerous state and national wildlife preserves. Moreover, opossums, skunks, deer, beavers, raccoons, wild turkeys *(Meleagris gallopavo)*, and large flocks of ducks and geese seem to have become increasingly adapted to the entrenched human populations.

The value of the forestland has been superbly demonstrated in this region. Beneath the mat of pitch pines, meadow grasses, cranberry bushes *(Viburnum trilobum)*, and other vegetation of the so-called New Jersey Pine Barrens, at the extreme northern end of the coastal plain, is an immense subterranean reservoir, the finest aquifer in the northeastern United States. It could supply a billion gallons of water a day to thirsty cities in the future without endangering the supply. Mature temperate hardwood forests of eastern North America produce a great deal of useful vegetative matter each year.

With such productivity it is little wonder that a complex,

dynamic cycle of living organisms thrives, even under fallen leaves. Copperheads *(Ancistrodon contortrix)*, those poisonous reptiles that bite more human beings than any other snake in the United States, blend effectively with the brown leaf patterns among which they live; their food supply is gained by pursuing small rodents, large insects, and caterpillars.

Insects on the ground and up in trees have formidable enemies in woodpeckers. The pileated *(Dryocopus pileatus)*, largest of the North American woodpeckers (unless the ivorybill is still alive, possibly in the Texas thicket), consumes ants and beetles as well as seeds, nuts, berries, and fruit, a varied diet which suggests that it has to live in rich dense forests. Happily, it is also returning in some numbers to places where it had been extirpated by extensive cutting of forest.

The Wild Turkey

The wild turkey, the largest of American game birds and common in much of warmer North America, is an able flier and supplies good eating. It spends much time in a wooded habitat; but because its newborn young are susceptible to chilling if they get wet from a summer shower, they must move to the meadow for warmth. Turkeys subsist on foods of the meadow, but to reach an average adult weight of up to 9 kilograms, they also need a substantial forest diet of nuts, berries, acorns, and miscellaneous items.

Their survival depends on success at a very weak link in the chain of life: they lay their eggs on the ground in either forest or meadow. This trait exposes the eggs and young to a host of avid predators, including raccoons, skunks, opossums, snakes, crows, and coyotes *(Canis latrans)*. But the adult turkey is capable of a fair defense, and may use deception to lead potential enemies away from the nest. Moreover, the clutch may contain as many as a dozen eggs. So the turkey has a better-than-even chance of survival. Most spend their nighttime hours roosting in the relative safety of trees.

The Flaming Woods of New England

Few colors are as brilliant as the autumn leaves of sugar maple *(Acer saccharum)* of northeastern North America. The coloration observed in leaves during the spring and summer months comes from chlorophyll, a green substance within plant cells. Most deciduous trees also contain red and yellow chemical compounds within their leaves, but these colors are masked during the growing season by the predominant green. Late summer and autumn bring cooler nights and days, decreased production of chlorophyll, and therefore a diminishing of the green pigments in broad-leaved trees. In this season the reds and yellows finally get a chance to show their true colors, providing a magnificent display of many hues of purple, scarlet, and yellow. Precipitation occurs frequently throughout the year, sometimes in cloudbursts of water and wind that lash the branches furiously. At other times not a leaf in the whole forest stirs. The aroma of rot and decay becomes familiar in this mixed deciduous-coniferous community.

Owing in part to a glacial alteration of drainage patterns,

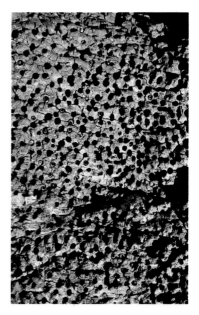

Few other birds are as well-equipped as woodpeckers to make effective use of woodland resources. All nest in cavities in trees, and most also feed on insects that live on trees or will eat fruits, nuts, and berries.
150 top. *A male red-bellied woodpecker* (Centurus carolinus), *sometimes called "zebra-back," consumes large quantities of wood-boring beetles, grass-hoppers, and other insects as well as nuts and wild fruits.*
Center. *The common flicker* (Colaptes auratus), *when feeding, seeks ants and beetle larvae on the ground but nests in tree cavities and drums loudly on dead limbs.*

150 bottom. *The redheaded wood-pecker* (Melanerpes erythro-cephalus), *which prefers open woodlands and orchards with dead trees, swoops after and eats many destructive insects. Found from eastern Canada to the Gulf of Mexico, it winters in southern South America.*
Above. *The acorn woodpecker* (Melanerpes formicivorus), *a colorful western bird of open oak and pine-oak forests, harvests nuts and packs them tightly into separate holes so that squirrels cannot dislodge them. Its storage trees include pine, Douglas fir, and oak; a colony of acorn woodpeckers may reuse the same holes for many years.*

Above. *Long claws, stout tails, and bear-like feet help yellow-haired porcupines* (Erethizon dorsatum) *climb trees in search of bark and other woody material to eat. This large rodent of temperate woodlands, seen here in Alaska, may be found from sea level to tree line and from deserts to rain forests.*

New England forests are broken by an abundance of ponds and lakes, some surrounded with oak, maple, or the peeling white-barked trunks of paper birch *(Betula papyrifera),* others covered with pondweed and lily pads. The rocks around them are covered with moss, which is slippery or spongy and makes our passage hazardous. The soft whitish reindeer moss *Cladonia* (actually a lichen) spreads across open vales in the forest. Such is the New England forest, much of which is now preserved and protected as national forests and parks.

Kingdom of the Grizzlies: The Boreal Forests

The Great North Woods stretch from Labrador to Alaska for some 6,400 kilometers. This is an immense boreal (northern) coniferous and deciduous forest that, in the Appalachian Mountains, survives as far south as Tennessee.

Although subject to winter weather extremes and to insects that devastate whole tracts, these woods managed to evolve certain adaptations and survival capabilities that have permitted them to develop rich ecosystems over the millennia. Three pines dominate: red *(Pinus resinosa);* eastern white *(P. strobus);* and jack *(P. banksiana).*

The boreal forest typically has two other major conifers, black and white spruces. The lesser of these is the black spruce *(Picea mariana),* often found in cold swamps, where it may grow very slowly: trees as little as 6 centimeters in diameter have been found to be more than 120 years old. The white spruce *(Picea glauca),* given good conditions of growth, grows up to 30 meters tall and to nearly a meter in diameter.

Much of the original forest habitat was slashed, burned, and severely razed; present-day forests, though second-growth, are recovering well where large preserves have been established. Wild coniferous-deciduous woods of the United States and Canada may be crossed for weeks at a time in several huge national and provincial parklands. In these wilderness areas the natural solitude, sounds, and beauty of the boreal forest are everywhere pervasive.

Like early *voyageurs,* we canoe along what seem to be lost waterways through endless woods. But the wall of coniferous trees is by no means all dark and gloomy; stands of white-barked quaking aspen *(Populus tremuloides)* mingle with needle-bearing trees, holding their own until eventually crowded out.

It should be remembered that some forms of wildlife, for example, the grizzly bear *(Ursus horribilis),* require a great deal of space. This great beast, at the apex of the American biological pyramid, seems to thrive best in regions allowing it at least 30 square kilometers of land per bear. Since that quota is not always available in its habitats, the animal has been progressively displaced to higher and more remote regions.

A bane of bears is the porcupine *(Erethizon dorsatum),* seven subspecies of which live in Mexico, Canada, and the United States. Once the sharp, barbed porcupine quills become embedded in the mouth of a bear—or any other animal—they can be painful and create such an impediment to eating that starvation may result. It is survival of the

fittest, and just as large animals have brute power to rip open small mammals, so the porcupines have an effective defense mechanism.

Wide bands of bark eaten from young, succulent pines and other trees are one indication that porcupines have been in the vicinity. The porcupine's adaptation to forests very likely came about because its ability to strip off tree bark and subsist on the soft cambium layer beneath sustained it through times when other foods such as herbs, catkins, and mistletoe *(Phoradendron flavescens)* were unavailable. It can also eat pine needles, which remain available throughout the coldest and snowiest of winters. Most of their peregrinations and feeding occur at night, thus accounting for regular confrontations with bears and other nocturnal prowlers.

Weighing 12 kilograms at most, these rodents may not be large, but they seldom have to worry about the size of their enemy when they defend themselves. Evolution has endowed them with a specialized kind of hair—those stiff, self-replacing, abundant (some 30,000) quills.

A more sinister enemy (apart from man) is the fisher *(Martes pennanti)*, a weasel-like animal that strikes the porcupine in its most vulnerable area—the soft underbelly. Fishers, sturdy fur-bearing dwellers of tree and ground habitats (but having little or no relationship to fish), are typical inhabitants of the boreal forest, ranging principally across Canada from the Atlantic to the Pacific and farther south along the major mountain ranges. No other tree-dwelling mammals are as nimble: fishers outrun squirrels, martens, and even snowshoe hares *(Lepus americanus)* on the ground. Such swift, persistent aggression has made fishers highly successful, for they will attack practically anything and eat nearly everything. They flip porcupines over on their back and rip into the vulnerable stomach area—perhaps picking up a few quills in the process.

The fisher (Martes pennanti) *inhabits forests in northern New England, New York, and much of Canada, where it preys on porcupines and other animals— sometimes even on deer floundering in deep snow.*

Wild Dogs: Coyotes and Wolves

Wherever you travel in the boreal forest of North America, it is very likely that you will be watched by a red fox *(Vulpes fulva)* or coyote *(Canis latrans)*. These ubiquitous wild dogs, which exert an effective control over the number of rodents such as ground squirrels and other small mammals in the forest ecosystem, are quite frequently misunderstood by many human hunters and farmers, who succeed in getting bounties placed on them. Nevertheless, against great odds, foxes and coyotes have survived the advance of civilization.

Less regularly seen is the gray wolf *(Canis lupus)*, even though it roams far and wide in the boreal forest and on the northern tundra. In their food-getting routine, wolves have a good weight advantage; they reach as much as 75 kilograms in more southerly habitats and 80 kilograms in the north. Another decided advantage in hunting prey is the capacity of wolves to work together. In so doing, they may attack in packs and exhaust large animals.

The Largest Mammals

The range of the moose *(Alces alces)* throughout the boreal forest roughly coincides with that of the wolf, though the wolf occupies regions much colder and more northerly than the moose. Both animals seem very much at home in open places such as meadows, but can be found in a variety of habitats. The moose frequents shallow ponds, lakes, and rivers, where willow and aquatic plants—its favorite foods—abound, but it is not averse to browsing on the vegetation of dry hillsides as well. Food-getting must be a continuous process for the moose, because it requires 18 to 27 kilograms of browse daily to nourish its bulk of up to 630 kilograms. As the largest living deer in the world, the moose presents a formidable challenge to enemies and is perfectly capable of stamping human intruders into the mud. The huge antlers, grown in late spring and summer, can serve as superb weapons—normally used in the autumn effort to win a mate. But moose are not invincible; grizzly bears can vanquish a full-grown adult moose. Young and weaker individuals are vulnerable to attack by wolf packs, especially in winter when the annual life cycle reaches a low point on account of severe weather or deep snow. Even if no wolves come around, the moose must be alert for coyotes, lynxes, and wildcats.

Moose roam the wild wooded parts of Canada. Not so the wood bison *(Bison bison athabascae)*, which is limited to a few preserves, notably Canada's immense Wood Buffalo National Park, the largest national park in North America. The bison there, one of the last remaining herds on this continent, are a hybrid of wood and plains bison. A herd of pure wood bison, the largest terrestrial mammal in North America, discovered in a remote corner of Wood Buffalo National Park in 1960 has encouraged hope that the subspecies can be sustained.

In summer wood bison feed mainly on leaves, shoots, and the bark of trees and shrubs; in winter they dig into the snow for lichens, dry grass, and moss. Watching bison paw the snow in icy winters or push it away with side-to-side movements of their ponderous heads, one would not think they could get enough sedges and grasses to eat. Indeed, sometimes they don't, and by late winter their ribs begin to show. But theirs has been a history of survival—except for their effective former hunter, man, who now instead strives to bring them back from the verge of extinction.

Home of Olympic Elk: Temperate Rain Forest

When 3,556 millimeters of precipitation fall on a forest each year, and the soil is fertile and the weather benign, one may easily expect a rich and luxuriant growth. Such is the case on the Olympic Peninsula of Washington state, especially in the Hoh River rain forest of Olympic National Park. Virtually everything in this dense growth of giant Sitka spruce *(Picea sitchensis)* and western hemlock *(Tsuga heterophylla)* is covered with sphagnum moss, a hundred species of mosses and liverworts, and 68 species of lichens. The principal lumber trees of the Pacific Northwest, Douglas fir *(Pseudotsuga menziesii)* and western red cedar *(Thuja plicata)*, rise up over an understory of vine maples *(Acer circinatum)*. Water lies everywhere in pools and springs, runs off in streamlets or rivers, and drips from dew condensed on leaves in the wake of passing clouds or the steady rain and drizzle.

The name moose (Alces alces), from an Indian word meaning "twig eater," describes the feeding habits of this largest of the world's living deer. The moose's range is now restricted to Canada, Maine, Minnesota, the Rocky Mountain states from Wyoming northward, and Alaska.

Above. A female moose is eating twigs and terminal buds in an Alaskan forest. Winter is a critical time for moose trying to find enough food to sustain their huge bulk; by the end of winter malnutrition may be common. Right. The bull moose, weighing up to 530 kilograms, has a distinctive large nose, throat dewlap, and an immense rack of palmate antlers. When rutting time approaches, moose remove the velvet from their antlers by rubbing them against trees, as do other deer.

These dim wet galleries, so difficult to penetrate, would hardly seem the place for mammalian life, especially large ruminants with antlers. Yet the Olympic elk *(Cervus canadensis roosevelti)* thrives as well in these soggy lowlands as on the slopes of the Cascade Mountains and somehow does not seem to have much difficulty getting through the interlocked tangles of red alder *(Alnus rubra)* and Pacific willow *(Salix lasiandra)*. For the most part, it feeds in meadows, but always with dense woods nearby. In fact, the elk themselves help to keep the meadows open; were it not for their browsing on the fresh green leaves and shoots of shrubs and seedling trees, the forest would be more dense than it is. By pruning the advancing vegetation, elk influence the growth of the forest. In September and October the woods resound with bellows, shrieks, bugling, and battles as the bull elk enter the mating season. Young elk are born the following June. Human visitors to these coastal rain forests seldom get a glimpse of the elk, for these large animals are usually quiet, unobtrusive, and shy. Moreover, once one is flushed, it is difficult to follow it and get a good look because of the density of forest growth. Yet some 5,000 elk inhabit Olympic National Park, and a persistent observer is likely to be rewarded with views not only of those animals but of such other wildlife of the rain forest as black-tailed deer *(Odocoileus hemionus)*, mountain lions and black bears. As for legendary animals such as "Bigfoot," or the sasquatch, we would reply as the Nepalis do when asked about their Himalayan yeti: Keep an open mind.

Land of Little Sticks: The Alaskan Taiga
Dense forest prevails northward along the coastal portions of British Columbia and southeastern Alaska. The Sitka spruce forest, sprung up after the melting of glaciers within the last 200 years, grows in a green hummocky environment replete with blue grouse *(Dendragapus obscurus)* and red squirrels *(Tamiasciurus hudsonicus)*. A thick woods so far north is explainable only by the mild influence of ocean currents off the Canadian and Alaskan coasts. No such mild oceanic influence moderates the climate of central Alaska; there the forest is reduced to a "land of little sticks," the small sparse spruce and willow stands of tundra and taiga. Barren ground caribou *(Rangifer arcticus)* wander in great herds through taiga forests and across gravelly plains. They are followed by wolves, which seek out weak members of the caribou herd, thus helping to maintain its health. The wolves are followed by eagles, foxes, and other carnivores, which raid the caches where wolves hide part of their kills. The huge moose may be encountered most often among willow patches or in ponds, where it dips its enormous head beneath the water to tear loose aquatic plants. Another giant creature of the Alaskan taiga, the grizzly bear, spends most of its time on meadows or tundra, digging up roots and ground squirrels. Birds nest in large numbers or congregate on ponds or, like the grouse, creep quietly through the undergrowth. These are the natural treasures of the original Alaska, one of the last large relatively undisturbed landscapes on earth.

The American elk, or wapiti (Cervus canadensis), ranks next to the moose in size (up to 450 kilograms); the male's antlers alone may weigh more than 20 kilograms. Elk now survive only in widely scattered, mostly isolated groupings, from coast to coast and from Texas to Canada.
160 top. A familiar sight to visitors in Yellowstone National Park, Wyoming, in late May or June: an elk mother recognizes her own calf by its odor and hides it in dense thickets to escape detection.
Bottom. Toothmarks on a quaking aspen are evidence that winter is a time of suffering and hardship for elk herds, when normal winter forage is covered by deep snows. By late winter many elk are in a weakened, starving condition.
162–163. Elk in the Yellowstone area migrate each spring from low-elevation winter ranges to higher summer ranges where new growth is plentiful.

Major Types of Forest Vegetation

 Montane

Coniferous

Mixed

Deciduous

Tropical Rain

Central and South America

The greatest tropical forest on earth extends from Central America south through the Amazon Basin to southern Brazil. It is not one forest but many: hot, humid lowland forest; montane forest; cloud forest; rain forest; and dry, almost desert communities. Often the vast expanse of densely packed trees is broken by mountain ranges, lakes, and broad savannas. Along the western edge of the continent, we follow the Andes cordillera into the southern temperate zone and enter great beech forests where condors fly and pumas prowl. From north to south it is a continent of surprises.

Iguanas and Toads: The Low and High Forests of Central America

To travel south into Mexico and Central America is not necessarily to enter strange new worlds. Just as in other parts of North America, there are pine groves, oak woods, mangroves, and deciduous and rain forests. White-tailed deer *(Odocoileus virginianus)* roam open places all the way to northern South America. Yet there are remarkable differences. The oak forests of Central America, for example, may be drenched with precipitation and support rich clusters of air plants. Arid zones there may be dry for six months and flooded for six months.

The variety of ecosystems in a single region of southern Costa Rica is impressive. The trees are no less imposing. The massive guanacaste *(Enterolobium cyclocarpum)*, though chosen as the Costa Rican national tree, is only one of 1,800 species of trees in that country. Even the undergrowth in wet tropical forests can be spectacular; the leaves of *Gunnera* can grow to 3 meters across. Flowering trees offer bright spots of varied color on the landscape, principally members of the genus *Tabebuia;* the sulfur-yellow flowers of *Tabebuia chrysantha* that burst out in April look like explosions of flame above the lowland forests. On a more subtle scale, orchids are seen nearly everywhere in tropical forests. A splendid example is the Spanish flag orchid *(Epidendrum radicans)*, with its brilliant red and yellow flowers—merely one of the thousand known wild species of orchids in Costa Rica alone.

The lowland forests may be dense and alive with howler monkeys *(Alouatta)*, whose echoing choruses fill the afternoons with waves of sound. Flocks of parrots fly overhead. Sally lightfoot crabs *(Grapsus grapsus)* scurry beneath palm leaves.

Nothing could be more startling than having a meter-long iguana *(Iguana iguana)* leap down and crash-land on the trail behind you in a tropical forest. That has happened to us more than once in a Costa Rican national park, where these reptiles are coming back after long persecution by human hunters who find iguana meat tasty. Although the iguana lays its eggs on beaches and may take to the water to avoid enemies, for most of its life it keeps to trees.

Its skin patterning usually blends with trees, and the iguana instinctively remains absolutely still when danger threatens. But if threatened, it can run and leap and get away rapidly. Iguanas are characteristic of the New World, but the fact that a few can be found in Madagascar and the

166–167. *Common iguanas*
(Iguana iguana) often sun
themselves on perches high in the
forest. Found in wooded country
from Mexico to Brazil, these
harmless reptiles run down tree
trunks and flee when danger
approaches, leaping off limbs onto
the ground or into water.

Fiji and Tonga Islands suggests they once lived in forests the world over.

In the less accessible uplands of Costa Rica, where moisture rises, condenses into clouds, and bathes the slopes and ridges, we find abundant vegetation and relatively undisturbed cloud forests. Here lives the richly colored quetzal *(Pharomachrus mocinno)*, a bird whose elegant red and green feathers the Mayans and Aztecs used ceremonially and for which the people of Guatemala named their currency. Now an endangered species, the quetzal is scarce and hard to find; after spending days in Central American cloud forests, we never observed more than its swooping shadow in the topmost branches of the forest canopy. These long-tailed birds, which nest in holes in decaying trunks, 4 to 20 meters above ground, eat fruits for the most part but occasionally take insects, frogs, and lizards.

One of the richest protected environments in Costa Rica, the Monteverde Cloud Forest Reserve in the Tilaran Cordillera, consists of six ecological communities that contain a total 2,000 species of plants, 320 bird species, and a hundred types of mammals. Among the latter are jaguars *(Felis onca)*, pumas *(Felis concolor)*, ocelots *(Felis pardalis)*, tapirs, peccaries, deer, and monkeys. Apart from quetzals, the bird population includes great green macaws *(Ara ambigua)* and black guans *(Chamaepetes unicolor)*.

Frogs and toads abound, and a brief search in virtually any Central American forest will disclose them, usually camouflaged, in their hiding places. Some frogs are exceptionally colorful and markedly patterned with stripes or spots; others are poisonous. In a very practical way, their poison protects the frogs, for potential enemies learn not to molest them.

Such extraordinary natural adaptations have evolved in frogs and toads to counteract their precarious situation, namely, that they are small, good eating and are surrounded by many predators. Like iguanas, most frogs depend on camouflage and patience for safety and also in getting their own morsels of food. They will burrow so far under the vegetation on the forest floor that only their eyes and noses protrude. Then at the precise moment their jaws snap open and shut like steel traps on an unsuspecting mouse, lizard, insect, or perhaps another unwary frog.

These tropical frogs and toads are not quiet animals. In Amazon forests, from a single listening post we have heard more than a dozen different chirps and warbles and squeaks. The Brazilian blacksmith frog *(Hyla faber)* is in fact named for its distinctive croaking, which sounds like the clank of hammer on anvil.

A favorite hiding place for frogs, toads, and other small inhabitants of tropical or subtropical forests is the leaf fold of epiphytes, or air plants. These vegetal growths obtain their nutrients and moisture from air and rain or mist, and they cling to tree limbs for support. They grow high in the treetops to gain the advantage of sunlight, and their colorful red, yellow, or green leaves add bright spots to the forest shadows. Many epiphytes belong to the pineapple

Due to the drenching humidity of rain forests, tropical tree frogs are exempt from aquatic life. Their adaptations to an arboreal environment are remarkable. 168 top. This male tree frog (Phyllomedusa trinitatis) in the Arima Valley of Trinidad has suction pads on its toes that assure a solid hold on the most slippery surfaces as it leaps from branch to branch. Bottom. Male and female tree frogs mating. In some species the female carries a cluster of eggs on her back until the tadpoles are completely developed.

Above. *Close-up of seven-day-old tree frog eggs, bound to the leaf in a gelatinous mass. Tadpoles develop in this aerial "nest" and, at seven or eight days, drop from their egg capsules into the water below.*

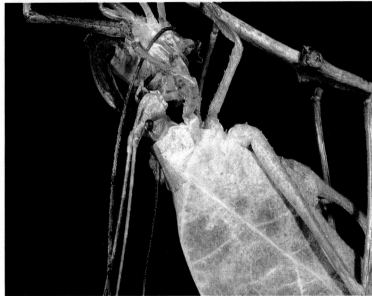

Among invertebrate life of Central American forests the battle for survival is universal. Katydids shed their skins in the moist night air and, like amphibians, must be alert for prowlers. Animals hide, set up defenses, or prepare to attack. Yet the age-old process of reproduction continues, and life, however dangerous, goes on.

Above left. *Bromeliad frog* (Amphodus auratus): *a close-up and a view in a drop of water in Trinidad.*
Above center. *A forest crab* (Pseudothelphusa richmondi) *seen in usual pose and in defense position at Barro Colorado, Canal Zone.*
Above right. *Katydid, in Trinidad, molting after having shed its skin; later, it eats the shed skin.*
Left. *A female land snail* (Strophochilus oblongus) *laying eggs on forest floor in Trinidad.*

family. Each is a microhabitat in itself, for tiny pools of water collect in cups formed where their leaves join the stems. To these dependable water sources, some organisms come to drink and others to live. There is hardly a niche in the tropical forest that is empty of life.

The Amazon Forest

The Macaya, a little lost river that hardly anyone knows about, flows into the Caquetá (Colombia), which flows into the Amazon. We remember the Macaya more than the big rivers of the Amazon Basin because it has not been hemmed in by *fincas*, those seemingly endless cultivated plots of farmland that line the larger rivers of South America. On the little Macaya, as we floated down the curving river in our dugout canoe, the noises of the forest were all around.

At first they were subtle noises, subdued distant howls and growls filtering through the trees. A pair of doves exchanged what seemed like mournful laments; insects buzzed; leaves rustled. Every sound and whisper seemed at first muffled by dense walls of tropical forest on both sides of the river. Vegetation grew so close to the waterside that seldom so much as a mudbank showed beneath the overhanging limbs whose leaves dipped into the water and drifted from side to side on the current. Giant *Morpho* butterflies, with iridescent purple wings up to 20 centimeters wide, flew above us in this vast "canyon" of lush foliage.

Suddenly a loud but melodic choking sound pierced the forest, a staccato bell-like burst so pervasive that it seemed to come from everywhere at once. For days as our canoe floated down the Orteguasa River, we had seen the pendant, almost bulbous nests of the crested oropendola (*Psarocolius decumanus*). This bird suspends its home from the outermost limbs of isolated trees in meadows or along streams, so that predatory mammals and reptiles have a difficult time reaching its eggs and young. Meticulously woven, the fiber suspension cords and nests, several meters long and sometimes 50 to a tree, sway with the slightest breeze or swing in dizzying arcs in a violent rainstorm. We had seen these black birds, with flashing yellow patches on the tail, fly in and out of their nests. But only here, finally, in this wilder stretch of forest did we hear their ringing bell-like song. If we could have observed them in their ritual mating antics, we might have seen the males raise their trembling wings and beat them together in display or perform cartwheels near the nests.

We wanted to see firsthand whether a tropical rain forest in the Amazon Basin was indeed as luxuriant as Charles Darwin had suggested in his writings. As we penetrated through the wall of greenery on the riverbank, we found the interior less dense than expected; there the mature forest remained open, relatively devoid of undergrowth. We shuffled through a carpeting of large thick leaves, but low green vegetation was scant among the dimly lighted trunks of giant trees. Clearly, one could find more luxuriant vegetation in temperate rain forests or even in tropical cloud forests. Nevertheless, these woods survive on solar

energy. The thunderous, deafening downpours of rain we experienced practically every night in the Amazon forest merely leach the minerals from the soils. The richness of these woods derives from the sun-drenched leaves and stems of their great trees—for example, the silk cotton *(Ceiba pentandra)* and species of *Bombax*—as well as the wild animals that thrive among them.

Land of the Jaguar

Tropical forest, thickets, scrubland, or rocky terrain all suit the jaguar, though its preference is for sheltered places. It climbs less easily than the African leopard *(Panthera pardus)*, perhaps because of its stockier build and somewhat larger size, up to 135 kilograms or more. But, agility aside, this cat is powerful: it can break the neck of a horse and sometimes attacks domestic animals. In wild forests it has to be opportunistic, however, for most of the animals it depends upon as food supply have developed a wariness that is usually effective. But peccaries *(Tayassu* spp.), fast though they can be in escaping through the woods, have become a staple of the jaguar diet. If the jaguar can find tapirs, capybaras *(Hydrochoerus hydrochaeris)*, or deer on the ground, so much the better. If it has to pursue food through the tree branches, it may catch a slow-moving sloth fairly easily, but will have little chance of keeping up with monkeys or birds unless it comes upon injured specimens or waits patiently in ambush. Because the jaguar swims well, not even turtles, fish, or amphibians are safe from its attacks.

Margays *(Felis wiedi)*, another wild species of cat, are pure forest dwellers and excellent climbers, capable of even running down huge tree trunks headfirst. In contrast to the ground-roaming jaguars, the much smaller margays hunt their prey in treetops.

On the ground live herds of fast-moving peccaries, wild pigs with long slim legs and delicate pointed toes. It is well to leave them alone, for though usually docile, they will counterattack when wounded. And if a whole band of as many as 100 peccaries attacks, the intruder could be in trouble. Peccaries prefer to live near rivers, and although they are good runners and travelers, they may never range beyond an area of 5 square kilometers in their lifetime. One reason for this confinement is that they will eat practically anything which enters their range or grows within it: fruit, berries, and bulbous roots mostly, but now and then a snake or some other small vertebrate. In turn, they are fair game for jaguars, and to avoid capture they need every ounce of strength and dexterity in those slender legs and delicate hooves.

The tropical forest birds sometimes present a dazzling display, for there are nearly 3,000 species of birds in South America, a spectacular variety. And very likely the first that a visitor will hear or see is one of the raucous, dazzling macaws: the scarlet *(Ara macao)*, the blue-and-yellow *(A. araraura)*, or any of a dozen more. These members of the parrot family will chatter by the hour in a tree or utter sharp shrieks while flying overhead.

The hummingbird, which has reached a high degree of specialization in the Western Hemisphere, zooms through

Playing a large role in flower pollination amid tropical rain forests are the hummingbirds (Trochilidae), which seem to be attracted not by scent but by bright colors, especially red and orange. Hummingbirds hover on rapidly beating wings to gather nectar; in doing so, they transfer pollen from one flower to the next.

Above. *Largely because nutrients are concentrated in the soil's top few centimeters in tropical rain forests, many trees do not send roots very deep into the soil. Because shallow root systems provide little support, trees like this one in the Amazon Basin develop prop roots to keep them upright.*

the forest with great activity and flashing color. There are 235 species in South America, developed into individuals with striking iridescent plumage, feather crests, and unusual beaks and tails. Their descriptive names alone give some idea of the remarkable variety of forms: for instance, black-tailed trainbearer *(Lesbia victoriae)*, sword-billed hummingbird *(Ensifera ensifera)*, and violet-tailed sylph *(Aglaiocercus coelestis)*. This immense family, smallest of birds, feeds on nectar and insects; the protein supplied by the latter is important to the maintenance of the high energy level needed to sustain the hummingbird's power of flight. These birds are the only pollinators of certain flowers, especially those with long bright orange or red tubular petals, conspicuous in daylight.

In tropical forests such as these, having little wind, few wind-pollinated plants have evolved, leaving a crucial role in the plant reproductive cycle to the uniquely adapted hummingbirds.

More sedate, but as colorful, are the toucans (family Ramphastidae), among the most characteristic bird families of the American tropics. They feed on fruits, which are likely to be abundant in the rich rain forests around them, and like parrots are among the noisier tree-dwelling birds. As is so frequent in South America, a considerable variety of them has evolved, including toucanets *(Aulacorhynchus)* and aracaris *(Pteroglossus)*.

Gliding along the rivers of the tropics, we may flush a hoatzin *(Opisthocomus hoazin)* from its shrubby perch. The loud sounds of woodpeckers, of which there are at least 54 kinds in South America, filter through the trees. Thrush-sized horneros *(Furnarius rufus)* build oven-shaped nests on tree limbs. All these examples suggest some of the amazing variety of South American forest birds.

Dazzling Butterflies and Dreaded Ants

The lepidopterous fauna of the neotropical region is generally considered the most outstanding on earth—not the least because of the color, brilliance, and variety of its forms. One family alone, the Erycinidae, contains more species than can be found anywhere else in the world. Moreover, a large number of butterflies that live in the tropical rain forest of South America can be seen nowhere else. Many of the most magnificent—*Morpho, Agrias, Troides*—inhabit the high leaf canopy and thus are difficult to observe in their native habitat.

But no matter how spectacular a display, butterflies were not colored simply to please the human eye. Their colors may provide them a measure of camouflage while resting on a flower; more significantly, color is used for species recognition. Butterflies also have their place in the forest ecology, which includes eating and utilizing plant material, spreading pollen from flower to flower, and being eaten by parasites, scavengers, and such predators as lizards, birds, bats, and monkeys. With this onslaught, it is fortunate that nature provides for the reproduction of a lot of butterflies. Naturalist-explorer Henry Walter Bates estimated that there were seven hundred species within an hour's walk of a certain town.

Adult female butterflies lay an abundance of eggs, after

Tropical American butterflies excel in color, brilliance, and variety. They play a major role in the forest's multiple food chains, using plant materials for growth and reproduction and, in turn, becoming food for other creatures. They usually inhabit different levels of the forest—some the shady undergrowth, others the treetops and open places.
Top row, left. A transparent, or glass-winged, butterfly of the family Ithomiidae in Costa Rica. Many members of this group are repulsive to predators because of their bad flavor. Center. A Morpho *butterfly of Brazil. Morphos are almost always at or above treetop level, except when drinking along riverbanks. Right. Mating* Thecla *butterflies, sometimes called hairstreaks, often lay their eggs in flowerbuds,* *where the larvae eat as the flower unfolds. Bottom row, left. A transparent* Callitaera *of the family Satyridae. When alarmed, many satyrids will dodge into thick shrubbery and fold their wings above their backs; their dull undersides then make them very difficult to detect. Center.* Morpho peleus-insularis. *The larvae of morpho butterflies live gregariously in a web spun over the leaves on which they feed. Right. The "eighty-eight" or "eighty-nine" butterfly (*Callicore*), so-called because of the configuration of its wing markings, which serve to frighten or draw attention from more vital body parts.*

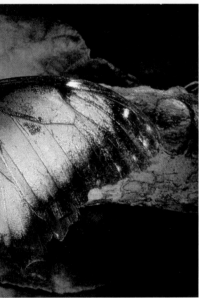

Left. *A large cluster of* Papilio astyalus *butterflies near Iguazú Falls, Argentina. Mass movements of butterflies or actual migrations take place with changes in climate and scarcity of preferred food plants.*

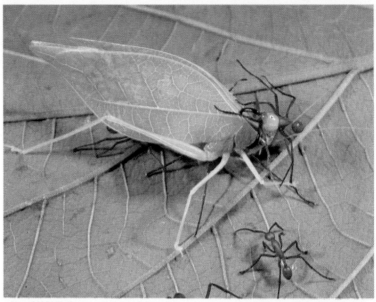

which the reproductive process enters three stages: larva (caterpillar), pupa (chrysalis), and adult. Mature butterflies have acute visual perception. Their senses of smell and touch are located in the antennae. Though generally solitary, butterflies may assemble in large numbers and travel to avoid drought and excessive rain or heat.

But it is their vivid, endlessly varied color that captures our attention. One of the most beautiful butterflies in the world, and one most sought after, is *Agrias sardanapalus*, which displays a rainbow of colors. The practical value of their colors and markings is demonstrated by *Caligo beltrao* of Brazil, which is blue, black, and yellow and has huge "eye spots" on the underside. When threatened, this butterfly suddenly opens its wings to frighten its enemy with an owl-like face on its wing surfaces.

South American forests are literally alive with millions of small creatures, and the biomass—or total living weight—of the invertebrates exceeds that of all the mammals. Whatever the case, the insect variety here is astonishing. Truly menacing insects are the army ants (*Eciton* spp.). These aggressive ants march through the forest in colonies of a million individuals at speeds of perhaps 20 meters an hour. Other animals flee from their path, for small creatures are seized in their mandibles and torn to pieces. Preying on this bonanza of ants are antbirds, mostly inconspicuous but constituting an enormous group of 226 species in South America. They move in bands, as they pick off the ants and the other insects disturbed by the passage of the marching horde.

South from the Tropics

Crossing to Paraguay, we stop briefly at the spectacular Iguazú Falls, 278 cataracts surrounded by subtropical trees, including ñandypá, guatambú, ysapuy guazú, araticú—a grand mixture, with names often taken from the language of the Guaraní Indians. There the little butterfly *Callicore*, with a distinctive "88" design on its wings, flutters about in open vales. Flocks of parrots circle the mists above the falls; hummingbirds fly through the woods; monkeys leap among the limbs; venomous serpents slither throughout. From deep in the woods comes the curious call of the chopí *(Gnorimopsar chopi)*, a blackbird whose notes sound like a comb being plucked.

Riding down a copper-colored dirt road on the approach to Iguazú Falls, we saw a small furry pig-like rodent cross the road and vanish into the thick green forest understory. Agoutis (*Dasyprocta* spp.), measuring up to 620 millimeters in length, burrow under tree roots and prowl in search of forest fruits and vegetables. But they are only one segment of the mammalian inhabitants and share their domain—warily in certain cases—with pumas, peccaries, tapirs, deer, and anteaters.

The Anteaters

Not only are certain birds adapted for hunting insects, but so are some mammals—none more so than New World anteaters, which dwell mostly in tropical forests. Serving their purpose well, the snouts of anteaters are extremely

A vast column of raiding army ants (Eciton) *is one of the most striking dramas in a tropical rain forest. Carrying their developing larvae along, these nomadic, predatory ants move nightly for 15 to 30 days, devouring any insects and small mammals they can catch. Then follows a stationary phase of several weeks, during which the queen, protected in a living nest of soldier and worker ants, lays thousands of eggs each day. When the larvae are about two weeks old, the colony moves on again.*

178 top. *Army ants discover a katydid.* Center left. *Close-up of an army ant.* Center right. *With large sickle-shaped mandibles, the ants tear off bits of the katydid.* Bottom. *Other ants join the feast.* Below. *Army ants mass to form a protective nest for the queen.*

Preying on ant and termite nests, anteaters (family Myrmecophagidae) generally inhabit tropical forests. Two of the three genera do most of their hunting in trees at night.
Above. *The profile of this immature giant anteater* (Myrmecophaga tridactyla), *displays the distinctive shape of its snout—elongated and tapered, with a tubular mouth that facilitates probing ant and termite nests.*
Right. *A lesser anteater* (M. tetradactyla) *feeds on termites in the lower Amazon Basin. Sharp claws are used to rip open hardwalled ant and termite nests. The animal compensates for its lack of teeth by a muscular gizzard that helps digest food.*

elongated and tapered, their mouths almost tubular, and their tongues long and sticky. Moreover, their forelimbs have five fingers equipped with long claws highly suitable for digging into ant mounds or nests. Giant anteaters *(Myrmecophaga tridactyla)*, hunting both by night and day, walk on their knuckles and rip open termite mounds. Collared anteaters *(Tamandua tetradactyla)*, which prefer to prowl at night, spend most of their waking hours in pursuit of ants, termites, and bees.

At a higher level of the forest, silky anteaters *(Cyclopes didactylus)*, which spend their nights in the treetops, use their long sticky tongues to catch ants and other insects, their deft movements aided by a prehensile tail and long claws. During the day they curl up in a hollow tree or on a forked limb. However, they must beware of enemies day and night, for in the treetops they are vulnerable to attack by eagles and owls.

It takes a great deal of ants and other food to keep these large animals going; giant anteaters weigh as much as 23 kilograms. Fortunately, the abundance of ants in tropic and subtropic ecosystems sustains populations of anteaters more or less in balance with the natural food supply. This delicate situation has prevailed for countless ages, disturbed only when human beings remove the forest to make way for farms, roads, and settlements.

Tapirs: "Living Fossils"

Traveling westward across wide fields and meadows dotted with ant mounds a meter high, we observe the spectacular purple flowers of the lapacho *(Tabebuia ipe)*, Paraguay's national tree. A highly useful tree, with wood so hard it would *quebrar el hacha* ("break the axe"), this became familiarly known as the quebracho *(Schinopsis lorentzii* and *S. santiqueño)*. Dry, open quebracho forests are found in the Gran Chaco, a huge wild region spread over parts of Paraguay, Bolivia, and Argentina.

The lowland tapir *(Tapirus terrestris)* is a mammal rarely seen by visitors to the forests of Paraguay, Colombia, Venezuela, and Brazil. During the Tertiary period of geologic history, rhinoceros-like species made up a large group of grazing and browsing mammals with divergent forms, but today only a few representatives remain: horses, rhinos, and tapirs. Called "living fossils" because of their more or less primitive physical structure, tapirs now live in widely separated regions. The Central American species *(Tapirus bairdi)*, weighing over 300 kilograms, is largest. Mountain tapirs *(T. pinchaque)*, weighing up to 250 kilograms and occupying the Andes of Colombia and Ecuador, are smallest. The more common lowland form *(T. terrestris)* weighs as much as 280 kilograms. The only other surviving tapirs live on the opposite side of the earth, in Thailand, Malaysia, and Sumatra.

The dark brown skin of tapirs, coarse and leathery, has little commercial value. Nor have these creatures an overpowering aesthetic appeal. Their upper lips have become almost trunk-like. Feeding on grass, aquatic plants, leaves, sprouts, and even small branches of trees, they seldom live a social life and have little means of aggression or defense. With jaguars around, they must

182. *Tapirs feed largely on leaves, fresh sprouts, and small branches they tear from shrubs and low trees. Their short elephant-like trunk is an extension of the upper lip. Most numerous is the lowland species* (Tapirus terrestris), *shown here in an Amazon rain forest, but all tapirs are now rare because of excessive hunting and forest destruction.*

always remain alert; human hunters also stalk them. When pursued, they can only hope to escape by crashing through the underbrush and eluding their pursuers.

Coihues and Arrayanes: The Southern Beech Forest

The genus *Nothofagus*, a name familiar to botanists, identifies 35 evergreen and deciduous species of impressive beech-like trees, some of them giants up to 40 meters tall, that grow in southern South America, New Zealand, and Australia. But the average Argentinian or Chilean will call them coihue *(N. dombeyi)*.

The forest strongly suggests a cathedral nave, with huge commanding columns—the trunks of coihues—sheltered by a parasol of lacy branches that let through slender shafts of sunlight to fall on rich banks of moss or tiers of lichen. The sound of carpinteros, Argentine woodpeckers, fills the woods with a staccato hammering. The most notable sound is the loud, sharp yet melodious songburst of the chucao *(Scelorchilus rubecula)*, a nondescript little bird rarely glimpsed in the deep shadows.

Unlike tropical forests, with hundreds of species of trees per hectare, the great southern beech forests more often have groves of single species scattered through the biome. Such is the distribution of one of the most attractive trees in all Argentina, the arrayán *(Myrceugenella apiculata)*, a member of the myrtle family. Its trunk exhibits various hues of red and orange, and when a shaft of northern sun enters the dark vale and falls upon a peeling, richly colored segment of bark, it creates a memorable image.

So do the blossoms of the notro *(Embothrium coccineum)*, a tree much smaller than the coihue but in a way more glamorous because of its spectacular clusters of scarlet flowers, which turn the tree into a blaze of color during spring (in November) in the Southern Hemisphere. So remarkable is this sight that Argentina has adopted the notro as its official national park flower.

Sloths: Primitive Tree Dwellers

Considering all the predators and other dangers in tropical forests, it would seem a logical outcome of evolution that each organism in such forests should be quick and nimble and that slow-moving mammals would have been eradicated long ago. But the slow-paced sloths are believed to be the most abundant and widespread arboreal mammals of the tropical forests of South America. We may not notice any in our progress through the woods; often these animals are too high up to be seen easily, or are well concealed by the dense vegetation. And that is one secret of their survival: camouflage. In fact, green algae growing on their fur helps them blend even better with their surroundings. Their extremely slow movements—which explains their name—make them virtually imperceptible. Nor need they travel far for food; the leaves and fruit they prefer, especially the common and delectable leaves of *Cecropia*, are all around them. With their claws hooked firmly onto tree branches, the sloths move along upside down, eating as they go and even sleeping in that position. And sleep is what they do most. Because their body temperature varies with the temperature of their surroundings, however, they

The primitive slow-motion sloths, completely arboreal in habits, can move on the ground only by dragging themselves along. Above. *The two-toed sloth* (Choloepus) *uses its curved, extended claws to travel upside-down along tree branches.* Opposite. *The three-toed sloth* (Bradypus tridactylus), *like its two-toed relatives, may center its life around a single tree.*

Few forms of life in Central and South American tropical forests are so strikingly "stratified" as the New World monkeys. Howler monkeys live in upper canopy layers, spider monkeys favor the middle layers, and capuchins occupy mainly lower stories and the forest floor. Marmosets prefer low scrub or forest edges, as do nocturnal monkeys, but there is little conflict because the latter seldom compete with daytime feeders. Top row, left. *The red-faced uakari* (Cacajao rubicundus) *is a slender, agile monkey whose range is limited to a very small area on the upper Amazon.* Center. *The lion-headed marmoset* (Leontideus rosalia), *so named for its golden mane, is an agile forest primate that usually feeds on spiders, insects, and fruit.* Right. *The night monkey, or douroucouli* (Aotus trivirgatus), *sleeps in pairs in tree hollows during the day, using its large nocturnal eyes to hunt for fruits, berries, tree snails, and insects by night.* Bottom row, left. *The hairy saki* (Pithecia monachus) *moves around in family groups, leaping from tree crown to tree crown in isolated areas of northern South America.* Center. *The mustached tamarin* (Saguinus mystax) *is a small diurnal marmoset that takes shelter in tree holes at night.* Right. *The douroucouli's coloration may vary, but only one species exists in tropical American forests from sea level to about 2,000 meters in altitude.*

are physiologically restricted to an equatorial habitat with minimal environmental changes.

Sloths belong to the same primitive animal order as ant-eaters and armadillos, all of them characterized by having poorly developed teeth or no teeth at all. *Choloepus* have two claws on their forelimbs, and the *Bradypus* three; hence their familiar names—two-toed sloth and three-toed sloth. So superbly adapted to its surroundings that it need do little more than eat, sleep, and move about—slowly and within a small compass—the sloth may be in a greater state of "ecological repose" than any other living mammal.

New World Monkeys: A Treetop Life

Before leaving the tropics and subtropics, we must review those most popular and abundant of South American mammals—the monkeys. Unlike the African or Asian monkeys, they are all arboreal. They also have long, well-developed tails, and most cannot oppose their thumbs to the remaining fingers to grasp an object.

Because broad rivers have formed barriers to the extensive mingling of species, there has been a great deal of speciation among New World monkeys. Squirrel monkeys (*Saimiri sciureus* and others), widely distributed in Central and South American tropical forests, feed on fruits, nuts, and berries as well as on any butterflies, beetles, and frogs they can catch. Capuchin monkeys (*Cebus* spp.) have similar patterns of existence. Howler monkeys *(Alouatta)* have so strong a tail that it is used almost as a fifth limb to hang by, climb with, and hold objects. They also have such a powerful voice that some people consider their sounds the loudest made by any animal. Alexander von Humboldt estimated that howlers' voices carried 2½ kilometers through the jungle.

Spider monkeys (*Ateles* spp.) possess the most highly developed prehensile tail. Old World gibbons depended so much on the use of their arms to travel through the trees that they lost their tails completely in the process of evolution, presumably because the tail impeded swift movement. In the spider monkey, the tail has developed into what amounts to another arm.

The high leaf canopy, a "forest above the forest" as Alexander von Humboldt called it, shelters the marmosets and tamarins (particularly those in the genus *Saguinus*), which look like a genetic mixture of cats, dogs, and monkeys. Some are tusked, some bearded; they are flat-faced, long-tailed, high-strung—in short an extraordinary bundle of energy and noise (bird-like and flute-like calls) in the tropical forest canopy. But in their chase for insects and search for fruit, such small animals must be quick and agile, and must be sure they don't inadvertently end up in the claws of a jaguar.

All monkeys of South America have a decided advantage over human visitors who try to get a glimpse of them in the wild. Their arboreal habits allow them to look down upon intruders and to disappear before visitors even know they are present. Since many are shy and seldom seen, information about their habits is scant. But in these New World ecosystems they are prime vegetarians and, as such, live an almost idyllic life amid rich tropical forests.

Woolly monkeys (Lagothrix), *which inhabit the rain forests of the central and upper Amazon Basin and Andean slopes, characteristically move cautiously among the tree branches.*

Arboreal Reptiles

Perhaps no part of the American forest ecosystem has inspired so much fear (and has been so misrepresented) as the reptiles, especially the big snakes. One sees dramatic setup scenes on television of grave riverside battles between men and anacondas *(Eunectes murinus)*, the world's longest snakes—up to 9.6 meters in length. In truth, however, the aquatic anacondas evolved to capture not human beings but wild forest mammals that came down to the shore for a drink.

In frightening terms we hear tall tales of boa constrictors 4 meters long, weighing 60 kilograms; but these more often inhabit upland forests where they feed on mammals, iguanas, and birds rather than human adventurers.

The emerald tree boa *(Corallus caninus)*, though beautiful to the human eye, uses its coloration as superb camouflage to help in its capture of prey. To many observers these are small miracles of an everyday ecosystem.

The Disappearing Rain Forest Heritage: *Patrimonio*

Today the places of civilization have grown larger and more numerous and have spread farther and farther into the countryside, thus raising the question: What of the future of South American forests, the largest on earth? Because of mounting population pressures and insatiable demand for wood products, charcoal, new roads and pastures, experts estimate that the seemingly vast tropical rain forests can disappear within relatively few years. Already, they note, some 40 percent of those which once existed in South America have been removed by man. The ecological relationships of plants and animals there are comparatively little known, but the flora is incredibly rich; New World tropical regions may contain as many as 90,000 species of flowering plants, 50,000 of fungi, and 5,000 of ferns.

Where tropical forests were cleared in Southeast Asia, precipitation patterns were altered, thereby producing heavier rainfall occurring at intervals longer than normal. This is perceived by some as an ominous indication of decreasing rainfall. Forest experts view the tropical forests as the world's most threatened ecosystem, and this dire prediction can be readily confirmed on a flight over them: the air is filled with great palls of smoke from woods being burned to clear land for agriculture.

Slowly, with the increasing emphasis on conservation, the tendency toward commercial exploitation of *all* animal life seems to be diminishing. And the more people learn of the treasures of tropical forests, the greater the demand for protection. Not long ago, government authorities in Colombia faced the task of preventing local settlers from moving into the Macarena Reserve, a wildlife-rich massif. In desperation they tried a new technique, which proved successful. "The Macarena is our *patrimonio*," they said in village lectures. "It is our heritage. If you disturb the forest and its wildlife, you will harm our country—the true Colombia so great in the eyes and minds and hearts of all of us." And, respectful of the unspoiled land, the people moved elsewhere, thereby saving another segment of the finest tropical rain forest in the world.

Picture Credits

Tree Evolution

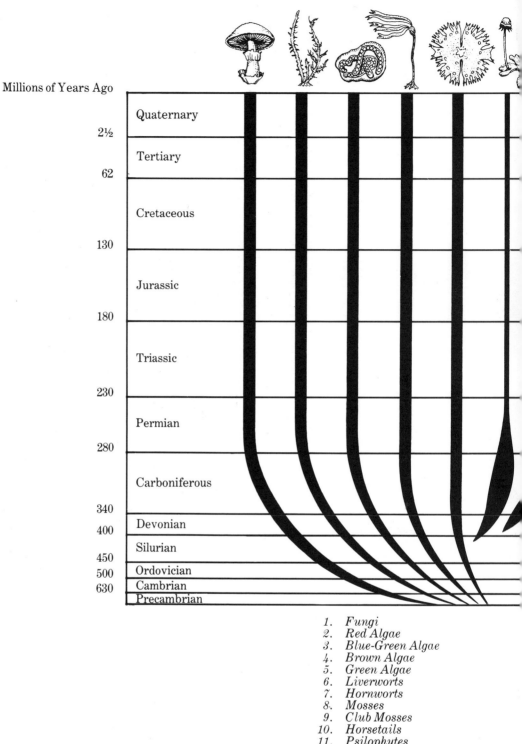

Millions of Years Ago

Quaternary	
	2½
Tertiary	
	62
Cretaceous	
	130
Jurassic	
	180
Triassic	
	230
Permian	
	280
Carboniferous	
	340
Devonian	400
Silurian	
	450
Ordovician	500
Cambrian	630
Precambrian	

1. *Fungi*
2. *Red Algae*
3. *Blue-Green Algae*
4. *Brown Algae*
5. *Green Algae*
6. *Liverworts*
7. *Hornworts*
8. *Mosses*
9. *Club Mosses*
10. *Horsetails*
11. *Psilophytes*
12. *Ferns*
13. *Cycads*
14. *Cordaites*
15. *Ginkgoes*
16. *Conifers*
17. *Seed Ferns*
18. *Flowering Plants*

The evolutionary relationships and the rise and fall of major plant groups are shown in this diagram. All land plants, including trees, evolved from green algae which were present some 2.8 billion years ago, and possibly even 3.1 billion years ago. Red algae, and their precursor brown algae, lived 900 million years ago. Fungi and blue-green algae are now generally excluded from the plant kingdom. Fungi are in the kingdom Fungi; blue-green algae are in the kingdom Monera.

The width of each life line on this chart indicates when and to what extent each plant type flourished. Thus, the club-moss group evolved early, developed rapidly, and was dominant for millions of years, but then became much less important in the world's vegetation. Seed ferns, once very common, are now extinct, but gave rise to most present-day land plants. Flowering plants, now everywhere and dominant, are latecomers, having appeared only about 150 million years ago.

Cambium layer

Inner bark

Heartwood

A tree is made up of cells, fibers, and channels that transport materials which give it life. Each year the growth of the tree adds a new sheath just under the outer tree bark. In cross section (above) these sheaths resemble rings. Movement of food, water, and nutrients takes place downward from the leaves and upward from the roots in the inner bark and soft layers called sapwood. As the tree enlarges, the cambium layer, just beneath the inner bark, produces new bark and sapwood, leaving a core of solid heartwood that gives the tree support. A leaf's cellular structure is shown in this magnified cross section (lower right). The food-making process of photosynthesis goes on in the leaf, which absorbs sunlight, carbon dioxide, and nutrients through its transparent surface cells and manufactures food in cells containing chloroplasts. By-products given off include oxygen and water vapor. Movement of food, water, and gases is carried on through cells and veins and leaf pores called stomata. Root tips like the cross section (upper right) probe the soil, multiplying by the thousands, and serving three principal functions: gathering water, collecting minerals from the soil, and forming a network that grips the soil and anchors the tree. Water and minerals are absorbed into root vessels. Most forest trees have their own root fungus systems, called mycorrhizae, which penetrate the root center, or cortex. They also feed on the tree's carbohydrates but give the roots increased access to soil nutrients.

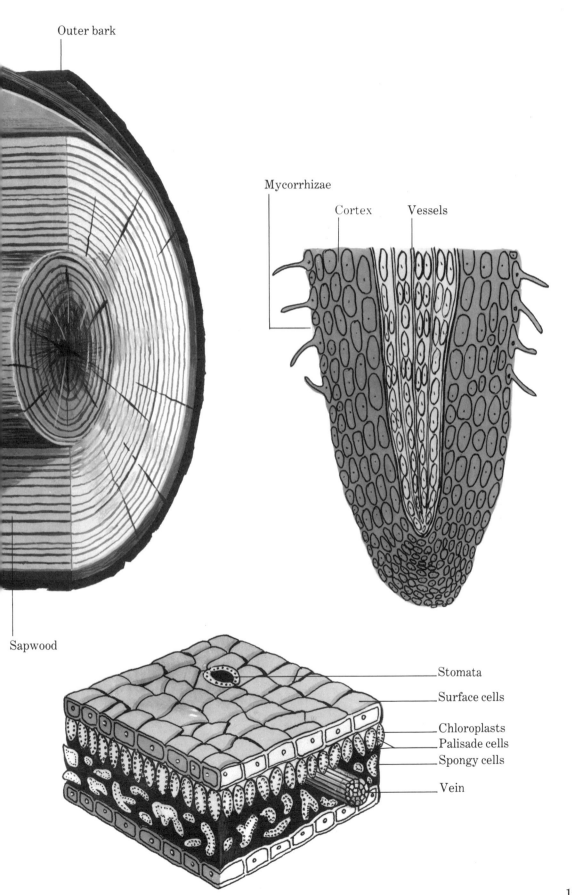

Outer bark

Mycorrhizae

Cortex

Vessels

Sapwood

Stomata

Surface cells

Chloroplasts

Palisade cells

Spongy cells

Vein

Principal Families of Trees

Taxaceae—Yew Family

One of the most widely distributed tree families is the yews, which include rather sparsely needled trees and shrubs. They can be found in Australia, New Zealand, the South Pacific islands, Africa, and South America and also grow abundantly in the Northern Hemisphere.

The yews proper (Taxus) *grow naturally only in the Northern Hemisphere and have great longevity; a few individuals have been known to live more than a thousand years. They possess flat, sharp-pointed needles; some with wide-spreading branches grow up to 30 meters in height and as much as 3,300 meters above sea level. Certain species may survive rigorous winters; instead of bearing cones, these produce small, plum-like fruits.*

English Yew *(Taxus baccata)*

Araucariaceae—Araucaria Family

These huge trees, perhaps the most ancient conifers, were once abundant in many parts of the world (sometimes found petrified in magnificent quartz), but now grow wild only in the Southern Hemisphere. The exceptionally symmetrical Norfolk Island "pine" is perhaps best known. Araucaria *provide a dramatic foreground for lakes and volcanoes among the Andean foothills of southern South America. Their sturdy, sharp-edged, pointed triangular leaves offer formidable resistance to any animal attempting to climb among them. Their rounded crowns, indicating maturity, may rise to more than 50 meters. Their female cones are often as large as a human head.*

Monkey Puzzle Tree *(Araucaria araucana)*

Cupressaceae—Cypress Family

Cypresses (Cupressus), *so familiar in mild climates, are graceful and aromatic, bear spherical cones, and possess a wood so durable that it can withstand even the continual assault of spray from the sea. Some cypresses grow tall in narrowly conical forms, while others assume broad, spreading shapes.*

Italian Cypress *(Cupressus sempervirens)*

Cypress Family (continued)
Junipers (Juniperus) vary from a small ground-hugging variety to large specimens with bark resembling alligator hide. Their fruits look more like berries than cones, and the leaves are either needles or scales.
The entire tree, including the wood when burned, is exceptionally aromatic.

Common Juniper *(Juniperus communis)*

Taxodiaceae—Redwood Family
The giant sequoias (Sequoiadendron) of the Sierra Nevada in California constitute the largest total bulk of any living thing. Though somewhat rare, confined to a mountainous strip measuring 12 by 160 kilometers at elevations up to 2,600 meters, they have become so well known and prized that their secluded groves are strictly protected in public parks. Thanks to a thick, durable bark, they endure natural fires quite well.
Equally protected by nature and man are the coast redwoods (Sequoia), which occupy a narrow lowland strip in the western United States. This stable, very humid environment stimulates these trees to grow well over 100 meters, making them the world's tallest.

Coast Redwood *(Sequoia sempervirens)*

Pinaceae—Pine Family
The pine family, including firs, pines, spruces, the most widely distributed of conifer groups, dominates much of Europe, Asia, and North America. Their hardiness is demonstrated by an ability to survive far north, where little else grows. Though stunted, members of this group make up taiga forest.
The family includes the true pines (Pinus) which tend toward temperate zones, where they inhabit very humid and dry environments. In both they sustain many life forms, including species such as the Kaibab squirrel (Sciurus aberti) that have become almost entirely dependent on them for food and shelter. Bristlecone pines (P. aristata) are estimated to be the oldest living trees, perhaps over 4,000 years old.

Western Yellow Pine *(Pinus ponderosa)*

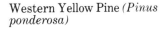

Pine Family (continued)

Spruces (Picea), *with stiff sharp needles and pendant cones, range much farther into the Arctic than do pines or firs. Though they grow to large dimensions along northern shores, where grouse, squirrels, and numerous other animals associate with them, they also cover vast areas as slow-growing constituents of cold taiga forests, where their roots may often be embedded in frozen soil.*

Norway Spruce *(Picea abies)*

Pine Family (continued)

Larches (Larix), *like a few other genera of this family, are deciduous, and their needles turn brilliant yellow in autumn before falling. Some larches in North America, Europe, and Asia grow to immense size.*

Japanese Larch *(Larix leptolepis)*

Pine Family (continued)

True cedars (Cedrus) *are survivors of trees that once formed a nearly uninterrupted forest through mountains from the western Mediterranean to the western Himalayas. Today they occur only in isolated patches: forests of Deodar cedar extend from the easternmost Himalayas to Nepal, and the famed rugged cedars of Lebanon have been reduced by woodcutters to a pitiful remnant.*

Deodar *(Cedrus deodara)*

Pine Family (continued)

The cones of hemlock (Tsuga) may be small but sometimes cover the ground in such large quantities as to form a light brown carpet. In moist and shady locations of Asia and North America, the hemlocks reach large size. Many can be readily identified by the drooping growth tip at the top of the tree.

Eastern Hemlock *(Tsuga canadensis)*

Salicaceae—Willow Family

This family, consisting of willows, poplars, and aspens, is virtually worldwide in distribution. There is an abundance of these trees along streams and in other moist places.
The willows (Salix) help stabilize natural ecosystems. Their widely spreading roots assist in holding soil in place along stream banks, thereby reducing erosion. These leafy shrubs and small trees are a principal habitat and food of the moose.

White Willow *(Salix alba)*

Willow Family (continued)

Most poplars (Populus) have grayish or whitish bark, with twigs and branches so brittle they are likely to break during heavy windstorms.
Willows and poplars may be cut by beavers for building dams. Poplars grow quickly, and their wood is soft. Aspens, other members of this family, have wide distribution and are able to survive at high elevations and in cold latitudes. Their bark helps to keep large animals from starving during severe winters.

Cottonwood *(Populus deltoides)*

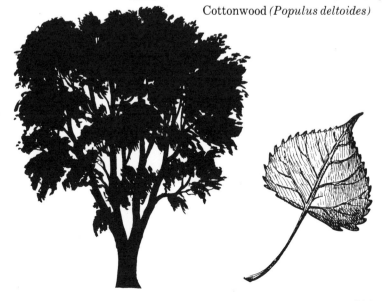

201

Juglandaceae—Walnut Family

Members of the walnut family are sturdy and slow-growing and have inconspicuous flowers. Some of the finest nut-producing trees, including hickories, pecans, and walnuts, they supply rich food for a host of forest animals.

Walnut trees (Juglans) *are found throughout Old and New World forests, where some specimens may grow to a height of 50 meters, with a trunk diameter of 2 meters. Though the nuts are solidly encased in hard shells, mammals such as squirrels are equipped to crack them and take advantage of their sustenance.*

European Walnut *(Juglans regia)*

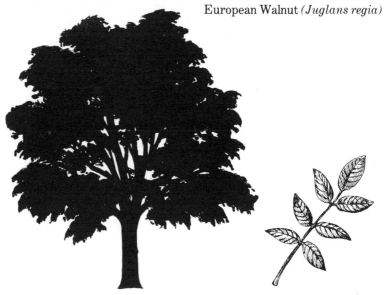

Betulaceae—Birch Family

Members of the birch family have simple leaves and tiny, inconspicuous flowers. Most are found over wide areas in temperate parts of the Northern Hemisphere, and some even survive in the tundra.

Alders, belonging to the genus Alnus, *are represented by 35 species of trees and shrubs. The genus includes the black alder* (A. glutinosa) *of Europe, which forms extensive forests from valley bottoms to elevations of more than a thousand meters across Europe into Siberia and south to North Africa.*

Black Alder *(Alnus glutinosa)*

Birch Family (continued)

Various birches (Betula) *are noted for their handsome, light-colored, sometimes papery bark. Many of the 60 species are found in Asia, but several are well known in North America. In Europe, the silver birch* (B. pendula) *and the downy, or European, white birch* (B. pubescens) *are both widespread. Some birches grow abundantly along stream courses winding through temperate forests.*

Canoe Birch *(Betula papyrifera)*

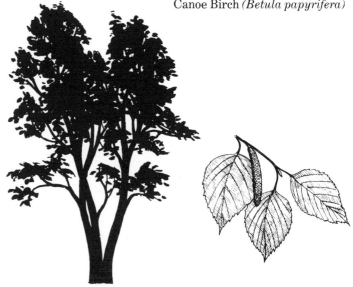

Birch Family (continued)
Hornbeams (Carpinus) *are slow-growing deciduous trees with some 26 species, found chiefly in Asia but with one* (C. caroliniana) *native of eastern North America. Others, including the European hornbeam* (C. betulus), *grow as handsome and imposing forest trees from western Europe to Asia Minor.*

European Hornbeam *(Carpinus betulus)*

Fagaceae—Beech Family
Members of the beech family, which may be either evergreen or deciduous, make up vast forests in the Northern and Southern Hemispheres. These trees produce large quantities of food each year in the form of nuts, such as chestnuts and acorns, which constitute a staple diet for many forest animals.
In the north temperate zone the beeches (Fagus) *produce small triangular nuts of value to birds, as well as dense shade and many surface roots. Owing to acid soil from leaf litter, few other plants can grow beneath beeches; this results in open forests when the trees mature.*

European Beech *(Fagus sylvatica)*

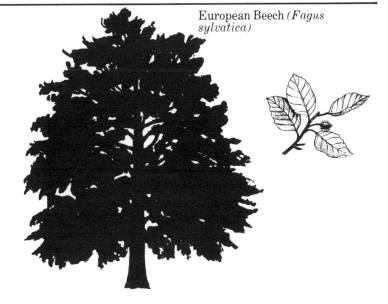

Beech Family (continued)
The durable oaks (Quercus) *have spread across North America, Europe, and Asia, ranging from the tropics to cold and icy environments. Drooping male catkins appear in early spring; then abundant acorns develop during summer from the female flowers. Oaks, too, provide wildlife with food, and like the beeches, their cast-off leaves are so acidic that very little will grow beneath them.*

White Oak *(Quercus alba)*

Ulmaceae—Elm Family

Trees of the elm family grow in warm temperate and tropical climates. In forests these huge trees may reach 50 meters in height; smaller varieties, such as hackberries (Celtis), often live along rivers.

Some varieties of elms of the genus Ulmus provide deep shade throughout the temperate zone of the Northern Hemisphere and adapt well to human settlements, where they have been extensively planted. Although some species of Ulmus are immune to fungus diseases—Dutch elm and phloem necrosis—others have been wiped out in great numbers because controls for these disorders have not been adequately developed.

American Elm (*Ulmus americana*)

Magnoliaceae—Magnolia Family

Among the most primitive of seed plants are the magnolias, which once grew throughout the Northern Hemisphere. Periods of glaciation wrought much havoc, however, and today their distribution is broken up into isolated stands in North America, Asia, and the tropics.

The huge white or pink showy petals of Magnolia grow at the base of a spiral central axis of stamens and pistils, contrasting with the dark green leathery leaves (the largest leaves and flowers in temperate forests). Insects gather the pollen of magnolias, and birds eat their seeds.

Southern Magnolia (*Magnolia grandiflora*)

Proteaceae—Protea Family

Great flowering spectacles are produced by this family which inhabits areas of southern Africa, South America, and Australia where long dry seasons are usual. Birds and insects are attracted to its flowers, which contain copious nectar. Banksia and Macadamia belong to this family as do Grevillea whose best-known member, the silky oak of Australia, has deep yellow flowers, which draw pollinators. The Embothrium trees of South America and Australia are known for their colorful displays of fiery scarlet or red blossoms, which grow on the branches in dense clusters and create the effect of a tree in flames—hence the familiar name "firebush."

Silky Oak (*Grevillea robusta*)

Lauraceae—Laurel Family

More than 2,000 species of trees and shrubs make up this tropical family, which is so numerous that not all the species are yet known. Included in this family of aromatic and poisonous plants are camphor, cinnamon, and sassafras trees.

The evergreen laurels (Laurus), which grow in the Mediterranean region and the Canary Islands, may reach 15 meters in height and bear inconspicuous flowers. Both species have aromatic leaves; one is called sweet bay.

Laurel *(Laurus nobilis)*

Hamamelidaceae—Witch Hazel Family

Trees of this family flourish in North America, Asia, Madagascar, southern Africa, and Australia; some reach moderate height. None survived in Europe after the Ice Age.

The witch hazels proper (Hamamelis) are unusual because they flower in autumn or midwinter. After ripening for a year, their seeds are literally shot out of capsules by a contracting and expelling process. In Asia some flowering witch hazels with fragile-looking petals are able to withstand temperatures below freezing without injury—a rarity among flowering deciduous trees in the temperate zones.

Common Witch Hazel *(Hamamelis virginiana)*

Leguminosae—Legume Family

This is a huge family of plants (12,000 species), some well-known members of which are pea, bean, locust (Robinia), the colorful mimosa, and more than 750 species of Acacia. Its fruits are characteristically borne in pods; roots develop nodules that contain nitrogen-fixing bacteria, which enrich the soil.

The widespread and abundant thorn trees found in arid regions are apt to belong to the genus Acacia. Their leaves are often small and feathery, yet provide some shade and moisture to animals in time of drought; their seeds and pods supply food. Especially in Australia, the flowers of some acacias produce exceptionally colorful displays.

Cassie *(Acacia farnesiana)*

Aquifoliaceae—Holly Family

Generally the holly family contains trees of separate sexes: each bears either male or female flowers, making "long distance" pollination a necessity before the female can reproduce. On occasion, however, bisexual trees and flowers may occur.

There are about 175 species of hollies (Ilex), *some of which are prickly leaved, some evergreen; some, but not all, produce red berries. Hollies grow in tropical as well as temperate zones, with some trees reaching a height of 40 meters. They may be found in such widely separated localities as the Himalayas and Brazil.*

English Holly *(Ilex aquifolium)*

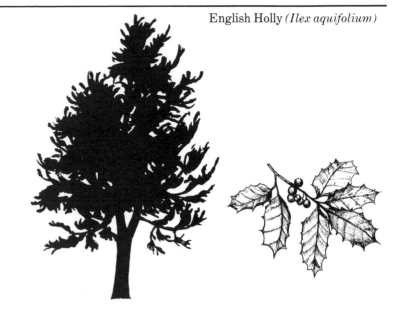

Aceraceae—Maple Family

The maple family consists of two genera and though its deciduous leaves are delicate and fragile, fossil imprints of them suggest that maples have been common and widespread for millions of years.

Some 200 maple species of the genus Acer *grow in Asia, Europe, and North America, flowering attractively in spring before the leaves appear and turning color in autumn to create a spectacular display. Several species produce a rich, sweet sap. Maples have become adapted to a wide variety of habitats, including stream banks, mountain slopes, and canyon walls.*

Sugar Maple *(Acer saccharum)*

Myrtaceae—Myrtle Family

This family comprises some 3,000 species of trees and shrubs, varying from small plants to a giant Eucalyptus *more than 100 meters in height. The tropical and subtropical Myrtaceae have elaborate-looking flowers consisting of numerous filamentous stamens.*

Most forest trees of Australia belong to the genus Eucalyptus, *of which there are some 500 species, many with attractive leaves and stems and colorful, patterned bark. Much wildlife depends on the eucalypts, and none more so than the highly specialized koalas; these slow-moving marsupials are totally dependent on the leaves of only a few species for food.*

River Red Gum *(Eucalyptus camaldulensis)*

Oleaceae—Olive Family

The olives sometimes live more than a thousand years, aided by a durable wood that resists external attack and decay. The small, inconspicuous flowers of Olea require twelve months to develop into the well-known dark olive fruit.

Olive *(Olea europaea)*

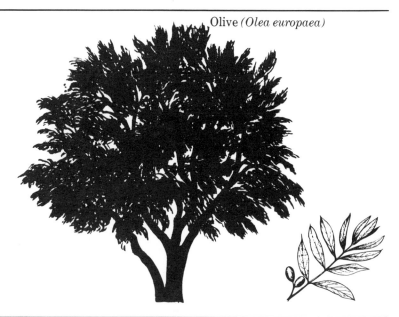

Olive Family (continued)

The flowers of ashes (Fraxinus), also members of this family, usually appear in clusters before their leaves grow, then develop into winged seeds that flutter on the wind and drift to the ground, thus spreading and perpetuating the species. Like those of maple and dogwood, ash leaves grow opposite each other on stems and branches.

European Ash *(Fraxinus excelsior)*

Palmaceae—Palm Family

The palms are not limited to the tropics, but most are found there—along seacoasts, in desert oases, or in mountain valleys. For wild birds such as the parrots and for mammals such as squirrels, they offer vitally important food, shelter, and nesting sites.
Although there are 3,500 species of palms in the world, perhaps the one best known to man is the coconut palm (Cocos nucifera), which ranges throughout the tropics and subtropics along seashores and for a short distance inland. Growing as tall as 30 meters, it bears up to 36 flexible, graceful leaf fronds. Its large fruits may float in seawater for months, eventually to take root on distant shores and thus expand the range of the species.

Coconut Palm *(Cocos nucifera)*

The Geological Record of Some Major Orders of Insects

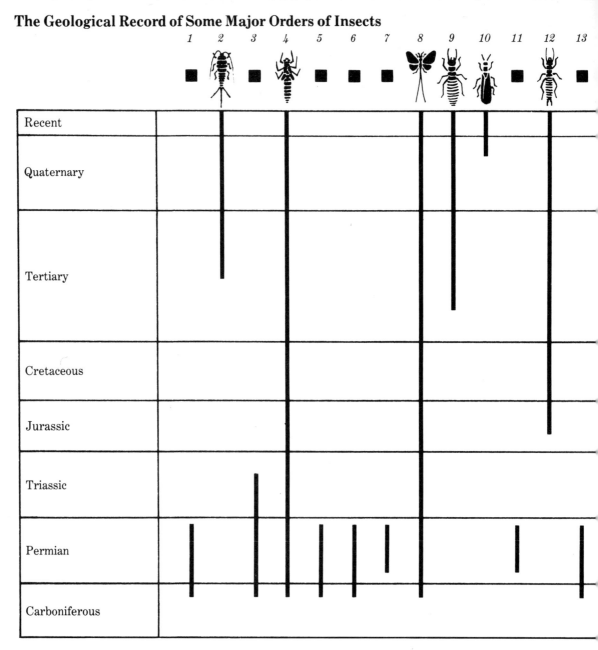

Palaeodictyoptera: Lithomantis carbonaria

Paraplecoptera: Lemmatophora typa

Protelytroptera: Protelytron permianum

The earliest fossils identified as insects date from the mid-Devonian period some 370 million years ago. Many insect fossils have been found in Illinois, Pennsylvania, and Colorado. Rich deposits occur in the limestone of central Kansas, in the shales of central France, and in Baltic amber. The early insects, like those at left, resemble springtails—members of the subclass Apterygota, which includes primitive wingless insects. The other subclass, Pterygota, includes insects whose thoracic structure is correlated with the development of wings; fossils of this group date from the Carboniferous period, some 45 million years later. Although many of these ancient insect groups are now extinct, the typical roach existed even then, as well as direct precursors of our present-day dragonflies. By the Jurassic period, many groups of insects had appeared in the basic form in which we know them today.

1. Monura
2. Thysanura
3. Meganisoptera
4. Odonata
5. Megasecoptera
6. Palaeodictyoptera
7. Archodonata
8. Ephemeroptera
9. Isoptera
10. Zoraptera
11. Protelytroptera
12. Dermaptera
13. Protoblattodea
14. Paraplecoptera
15. Protorthoptera
16. Plecoptera
17. Orthoptera
18. Embioptera
19. Miomoptera
20. Psocoptera
21. Mallophaga
22. Hemiptera
23. Thysanoptera
24. Neuroptera
25. Coleoptera
26. Strepsiptera
27. Mecoptera
28. Siphonaptera
29. Diptera
30. Trichoptera
31. Lepidoptera
32. Hymenoptera
■ Extinct Insects

Anatomy of an Insect

The class Insecta contains over 700,000 known species, more than any other group of animals in the world today. Insects generally reproduce sexually, and except for the small, primitive subclass Apterygota, all undergo a process of maturation known as metamorphosis, which may be either simple or complete. Complete metamorphosis has four stages: (1) the egg, (2) the larva, during which it feeds and grows, (3) the pupa, an inactive transitional stage, and (4) the adult, or imago. Metamorphosis is considered simple if the pupa of these stages is omitted.

The skeletal structure of adult insects is external: they are protected by a chitonous armor called an exoskeleton. Their bodies are usually composed of 20 segments, which over the millennia have fused into three defined body parts: head, thorax, and abdomen. The head consists of six fused segments; the first three bear the ocelli, or simple eyes, and the compound eyes and antennae used for touch, taste, and smell. The second three segments bear the mouthparts, which include mandibles (or sidewise moving jaws) and maxillae that serve for tasting and picking up food. To the three segments of the thorax (prothorax, mesothorax, and metathorax) are attached three pairs of segmented legs consisting of five divisions: coxa, trochanter, femur, tibia, and tarsus. In winged forms, two pairs of wings are outgrowths of the last two segments of the thorax. The abdomen, with 11 segments, contains the digestive and reproductive organs, or genitalia. At the end of its body, the female of many species has an egg-laying tube, called an ovipositor. Near the tip of the abdomen some primitive insects have an appendage, the cercus, having a sensory function. Insects have open circulatory and respiratory systems; air enters the body through spiracles, small openings on the sides of the thorax and abdomen. Some insects such as grasshoppers have an organ of hearing—the tympanum—located on the front legs, thorax, or abdomen.

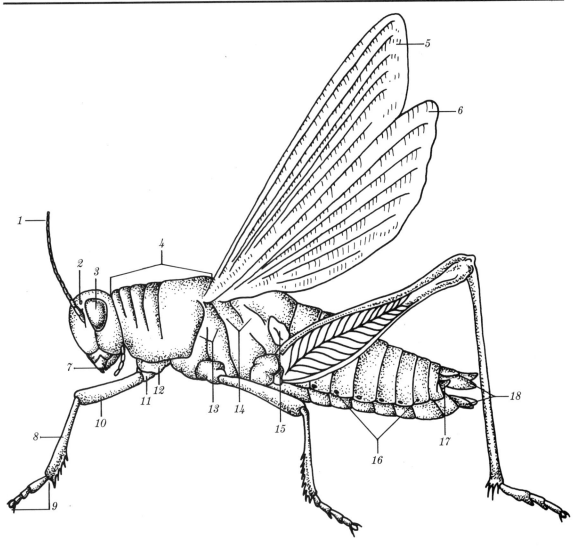

Order Orthoptera: Grasshopper

1. Antenna
2. Ocellus
3. Compound eye
4. Prothorax
5. Forewing
6. Hind wing
7. Mandible and maxillae
8. Tibia
9. Tarsus
10. Femur
11. Trochanter
12. Coxa
13. Mesothorax
14. Metathorax
15. Tympanum
16. Spiracles
17. Cercus
18. Genitalia

ORDER PROTURA — Proturans

Known only since 1907, the 118 species of proturans are soft-bodied, minute whitish insects measuring 0.6–1.5 mm long. They have cone-shaped heads, lack eyes and antennae, and carry a pair of appendages on each of the first three abdominal segments. Their mouthparts are of the piercing and sucking type. Their metamorphosis is simple and includes a process very rare in insects, though common in such primitive arthropods as milli-pedes: an increase in the number of abdominal segments from 9 to 12 during the three molts that occur in the larval stage.

Proturan *(Acerentulus barberi barberi)*

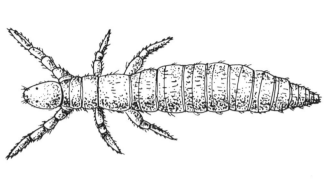

ORDER THYSANURA— Bristletails

These primitive, mostly neutral-colored, wingless insects, seldom exceeding 10 mm in length, take their common name from the three slender appendages extend-ing like tails from beneath the abdomen. A few of the 370 known species are familiar household pests; these include silverfish and firebrats, which feed on the starches in books, linens, and wallpaper paste. They display a strong avoidance of light but are attracted to warm spots. Like springtails and proturans, bristletails undergo virtually no meta-morphosis.

Firebrat *(Thermobia domestica)*

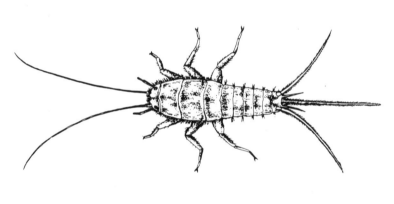

ORDER DIPLURA— Diplurans

There are approximately 400 species of these small wingless, eyeless insects, which live under bark, stones, soil, and debris and in rotting wood. They are similar in many ways to the Thysanura, but lack scales and have only two abdominal appendages protruding behind. They have concealed mouth-parts and are generally unpig-mented. Their metamorphosis is simple, with the nymphs closely resembling the adults.

Dipluran *(Campodea folsomi)*

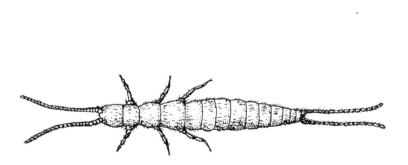

ORDER COLLEMBOLA—Springtails

With more than 2,000 species, these minute insects comprise the largest of the orders of primitive insects and occur abundantly in moist habitats throughout the world, including polar regions. Lacking wings, they nevertheless can jump many times their own length by means of a forked structure at the rear of the abdomen that acts as a kind of high-tension spring. The partially fused segments of the thorax are distinctive, and mouthparts are usually retractible. There is virtually no metamorphosis; the only difference between larvae and adults is the smaller size and sexual inactivity of the larvae.

Springtail *(Isotoma palustris)*

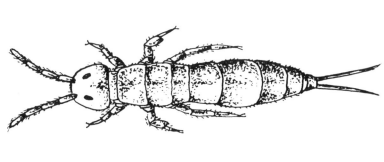

ORDER EPHEMEROPTERA—Mayflies

Adult mayflies are as ephemeral as their scientific name implies—living only a few days or, sometimes, less than a day. Their two pairs of wings, fragile and lacy, are needed for only a very brief time. During this time most species swarm in vast numbers and mate in "nuptial flights." The females then lay their eggs in fresh water, and the adult life cycle is completed. Mouthparts are vestigial or lacking, since the adults do not feed. Larvae (or nymphs), however, have plumose gills and live in water on aquatic plants for one to four years, finally emerging onto land in a preadult stage (subimago) unique to this order. Metamorphosis for the 2,000 known species is considered simple.

Mayfly *(Hexagenia bilineata)*

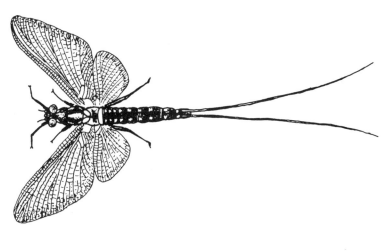

ORDER ODONATA—Dragonflies and Damselflies

Aided by a freely rotating head and huge compound eyes, these agile insects—comprising an order of 4,870 known species—are highly predatory, both as terrestrial adults and as aquatic nymphs. In the larval stage, they are an important link in the food chain, as both predators and prey. They have two pairs of elongated membranous wings: dragonflies hold their wings outstretched; damselflies close them above their long, slender bodies. Their metamorphosis is simple; the nymphs of both groups are aquatic, breathing by means of gills, and swim by undulating their external gills (damselfly) or by jetting water out their rectum (dragonfly).

Dragonfly Nymph *(Aeshna verticalis)*

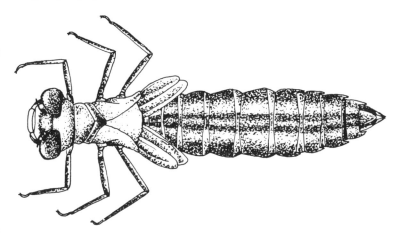

ORDER ORTHOPTERA— Grasshoppers, Walking Sticks, Crickets, Mantids, Cockroaches, and Rock Crawlers

This order is responsible for most of the "insect music" we hear; their sounds are produced by friction created by rubbing their legs or wings together. Nearly all 22,500 known species of Orthoptera are winged; the hind pair is folded fan-like under the leathery forewings. Mouthparts are of the primitive biting variety.

Walking Stick *(Carailsius morosus)*

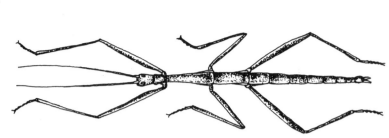

Order Orthoptera (continued)

Although some species of Orthoptera are predatory, most are plant feeders and, among these, certain migratory locusts may form immense swarms that can devastate croplands with incredible speed and thoroughness. Other than the polar regions, the Orthoptera occur worldwide. Their metamorphosis is simple; generally, larvae closely resemble adults, except for lacking mature sex organs and wings.

Field Cricket *(Liogryllus campestris)*

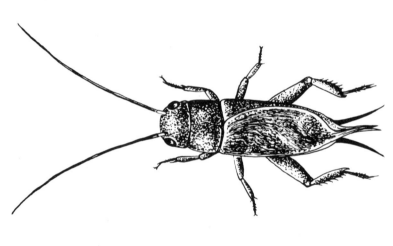

ORDER ISOPTERA—Termites

Some 2,100 known species of these small, primarily tropical insects live in large colonies, remarkable for the great differentiation among individuals and their sophisticated social organization. Typically white and soft-bodied, they have wing pairs of equal length and abdomens broadly joined to thoraxes. Both beneficial and destructive, they feed primarily on wood; the normally indigestible cellulose is processed by tiny protozoans living in their digestive tracts. Mouthparts are generally of a primitive biting type. Their metamorphosis is simple; the nymphs mature into workers, soldiers, reproductive understudies to the colony's king or queen, or new leaders themselves.

Worker Termite *(Prorhinotermes simplex)*

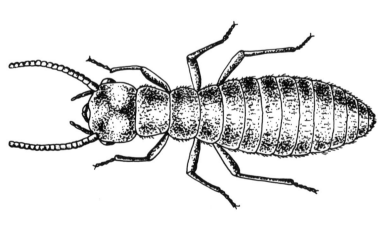

ORDER DERMAPTERA—
Earwigs

Earwigs are leathery, medium-size, flattened insects that number some 1,100 known species. Nocturnal in habit, they are primarily scavengers. Large abdominal, forceps-like pincers are used mainly for defensive purposes; mouthparts are of the simple biting type. There are some wingless species. Their metamorphosis is simple; in some species, females protect their eggs in a nest cavity and then brood and watch over the nymphs, which are quite active at birth. Such maternal care of the eggs and larvae is rare in the insect world.

Forest Earwig *(Chelidurella acanthopygia)*

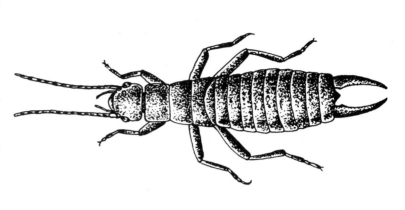

ORDER EMBIOPTERA—
Webspinners

There are fewer than 100 known species of these fragile insects, which spin and inhabit silken tunnels, generally in debris, on or under bark and rocks, and among mosses and lichens. Unlike most "spinning" insects, which produce silk from modified salivary glands, webspinners emit their sticky fluid from glands at the base of the forelegs. They are small (4–7 mm), and only the males may have wings on their slender bodies, hinged to facilitate passage through their tunnel habitats. Their metamorphosis is simple. Though primarily a tropical order, a few species are found in temperate zones.

Webspinner *(Oligotoma saundersii)*

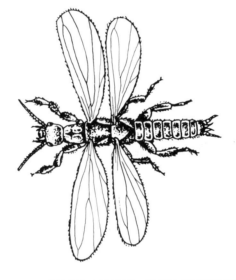

ORDER PLECOPTERA—
Stoneflies

Like mayflies, the 1,550 known species of this order while in the larval stage provide an important food source for freshwater fish. The presence of the larvae, found under stones in clear fresh water, is a fairly reliable indicator of water purity. Their eggs are dropped in water in loose packets. The elongated, flattened nymphs have gills on the thorax and at the base of the legs. The short-lived adult stoneflies are small to medium-size, with soft bodies, still rather flat and elongate. Their hind wings are folded fan-like under the long membranous forewings. Most species are weak fliers and remain near their home waters. Their metamorphosis is simple.

Stonefly *(Isoperla confusa)*

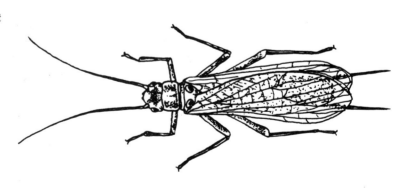

ORDER PSOCOPTERA— Psocids

These tiny insects (1–10 mm) have relatively large heads and soft bodies. Winged and wingless individuals may be found in the same species; when present, the wings are held roof-like over the abdomen. Their mouthparts are of the biting sort. The more than 1,700 species are of no great economic significance, though they may infest houses and feed on cereals, molds, and other materials found there. The misleading term "booklice" is applied to some species that feed on paste, glue, and paper in old books, but these are neither parasitic nor louse-like in appearance. The metamorphosis is simple.

Psocid *(Caecilius manteri)*

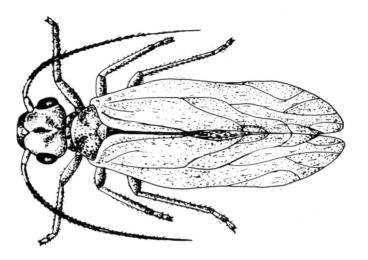

ORDER ZORAPTERA— Zorapterans

The 22 known species of these minute insects (less than 3 mm long) belong to a small order of gregarious, termite-like insects. Living under forest debris, under tree bark, or in termite nests, they feed on dead insects and fungus spores. Their metamorphosis is simple; the young live with the adults, and some mature into winged forms that eventually shed their wings.

Zorapteran *(Zorotypus brasiliensis)*

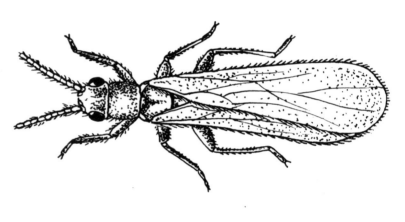

ORDER MALLOPHAGA— Chewing Lice

Although some of the 2,700 known species of these tiny, flat, wingless insects are external parasites on mammals, most exist as parasites on birds. Usually specific in their choice of hosts, they spread from bird to bird only when there is actual physical contact, as in the nest. Although not known to attack man, they are serious pests of domestic animals, infesting poultry especially. Their metamorphosis is simple, lasting only a few weeks. Except for size, larvae and adults are very similar and even eat the same food (hair, feathers, skin wastes).

Chewing Lice *(Menopon gallinae)*

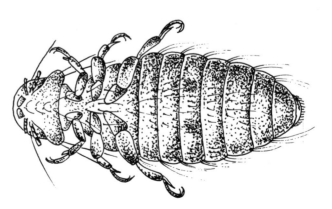

ORDER ANOPLURA—Sucking Lice

The 250 known species of Anoplura are parasites living only on mammals. Several species live on domestic animals, and man is the natural host to some species. Measuring less than 4 mm long, these tiny insects are, at best, sources of irritation to their hosts; at worst, they are carriers of dangerous diseases. Their very flattened, wingless bodies make them difficult to grasp and remove; their mouthparts are adapted for piercing the host's skin and sucking its blood. Metamorphosis is simple, with larvae and adults feeding on the same host.

Human Lice *(Pediculus humanus)*

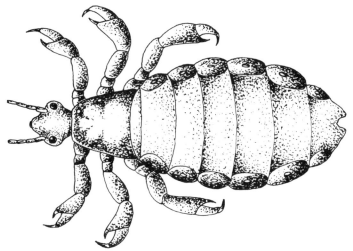

ORDER THYSANOPTERA— Thrips

This order includes some 4,500 species of slender, minute insects (0.5–5 mm long), ranging in tone from pale to very dark. They are chiefly noted as prolific plant pests and, as such, pose a great economic threat to agriculture. Wings are often lacking; when present, they are generally four in number and are long, thin, and fringed with long hairs. The mouthparts are adapted for piercing and sucking. The metamorphosis falls somewhere between complete and simple, with two larval stages followed by two or three inactive preadult stages; the last nonfeeding stage is sometimes spent in a cocoon.

Gladiolus Thrip *(Taeniothrips simplex)*

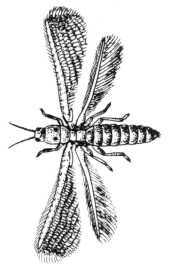

ORDER HEMIPTERA—Bugs

The term "bug" is popularly used to describe insects in general, but the true bugs, numbering some 23,000 known species, belong to this order. Varying widely in size, they are characterized by piercing and sucking mouthparts, and by forewings which are leathery at the base but not at the tip and which lie folded flat across the back and provide protection for the thin hindwings. The beaks that sheath the sharp mouthparts begin in front of the eyes. Species that feed on saps and plant juices present a menace for agriculture. Many members of this order are aquatic but lack gills; their air supply is trapped in the fine hairs covering their bodies. The metamorphosis is simple.

Stink Bug *(Thyanta custator)*

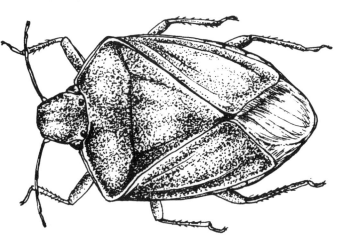

ORDER HOMOPTERA—
Cicadas, Psyllids, Aphids, Whiteflies, Hoppers, and Scale Insects

This diverse group of some 32,000 known species includes all insects with beaks attached behind the eyes. Wings, when present, are uniformly membranous and held roof-like over the back; but many members of the order are wingless.

Green Peach Aphid *(Myzus persicae)*

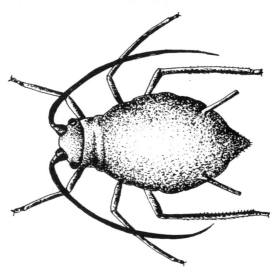

Order Homoptera (continued)

A common feature of Homoptera is their plant diet, shared by adults and larvae alike. Though a few species produce useful secretions, the great majority are pests, breeding in vast numbers, sucking vital nutrients from plants, and sometimes transmitting plant diseases. Their metamorphosis is simple.

Froghopper *(Philaenus spumarius)*

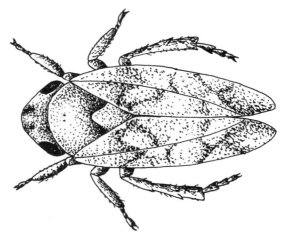

ORDER COLEOPTERA—
Beetles

Some 20 percent of all the existing plant and animal species in the world are beetles, and the estimated 290,000 species of Coleoptera comprise 40 percent of all known insect species. Beetles, varying in size from less than 1 mm to over 15 cm, are characterized by biting or chewing mouthparts and distinctive forewings (elytra) that have been modified into hard protective covers for the membranous hindwings. Metamorphosis is complete, with a high degree of specialization. Larvae are variable in form, body hardness, and appendage development. Like adult beetles, the larvae vary in feeding habits as well, from predaceous or herbivorous to scavengers and even parasites.

Ladybird Beetle *(Hippodamia convergens)*

218

ORDER STREPSIPTERA—Twisted-winged Parasites

This small order, numbering about 400 known species, is comprised chiefly of minute insects that live as internal parasites on other insects. Male and female Strepsiptera are very different: males have fore-wings reduced to club-like structures and broad, fan-like hindwings; females are wing-less, generally legless, and spend their entire lives in the host where they developed. Meta-morphosis is complete, but unusual in that it has two larval stages, a process known as hypermetamorphosis. During the first, active phase, the larvae emerge from eggs laid in the host's body and seek a new host. They then molt and assume a legless form and live on the new host.

Twisted-winged Parasite *(Neostylops shannoni)*

ORDER MECOPTERA—Scorpion Flies

This small but ancient order of insects takes its name from the recurved and bulbous genital structure of the male, which resembles the sting of a scorpion. There are only some 400 known species, which vary in size from 2 to 25 mm and which are harmless and mostly of little economic significance. A distinctive characteristic of this slender, soft-bodied insect is its curved beak. The metamorphosis is com-plete; larvae, like the adults, are generally predatory but are sometimes scavengers. The elongated hindlegs are used by the scorpion fly to seize other insect prey as it dangles from branches.

Common Scorpionfly *(Panorpa communis)*

ORDER NEUROPTERA—Alderflies, Dobson Flies, Fish-flies, Snakeflies, Lacewings, Ant Lions, and Owlflies

This order is made up of some 4,670 known species characterized by their four large wings and generally unsteady flight pat-terns. Found in a great variety of habitats, from deserts to moun-tain streams, they vary consider-ably in size. Their metamorphosis is complete, and the larvae are predaceous on other insects, sometimes beneficially if their prey is harmful to other organ-isms. They are important in natural food chains and eco-systems.

Ant Lion *(Dendroleon obsoletum)*

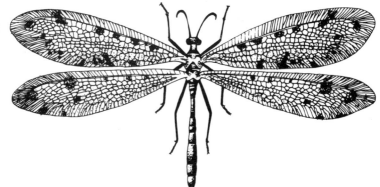

ORDER TRICHOPTERA— Caddisflies

Of the four insect orders that are aquatic in their early stages, caddisflies—numbering about 4,500 known species—are the only ones which undergo complete metamorphosis. The small to medium-size adults measure 1.5–25 mm in length, have slender, elongate bodies, and hold their wings roof-like above them when resting. The aquatic larvae live in cases they build from debris. Some spin nets in the water that are designed to catch small arthropods. The pupae crawl out of the water, and the adults emerge on land. They are an important food source for freshwater fish.

Caddisfly *(Macronemum zebriatum)*

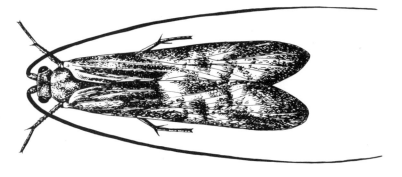

ORDER LEPIDOPTERA— Butterflies and Moths

There are over 112,000 known species in this large order, which includes some of the world's most beautiful insects. It is generally characterized by an adult stage in which the insect has four membranous wings covered with tiny scales and by the sucking mouthparts found in most adults, which feed by means of a coiled tube called a proboscis.

Monarch Butterfly *(Danaus plexippus)*

Order Lepidoptera (continued)
Butterflies, which tend to be active during the day, hold their wings erect and together above their backs when at rest. In contrast, the moths, which are generally nocturnal, rest with wings held roof-like over their bodies. Metamorphosis is complete, with a pupal stage during which the insect is transformed from a plant-feeding caterpillar into a winged adult. Larvae can be very damaging to plant life, whereas the adult is a valuable pollinator.

Luna Moth *(Actia luna)*

ORDER DIPTERA—Flies

This large and diverse order numbers over 90,000 known species, occurring worldwide. Its characteristic feature is the possession of only one pair of wings, the hindwings having been modified to a pair of long-stemmed knobs used to maintain balance. Diptera feed only on liquids; some, such as mosquitoes and horseflies, prefer blood and, along with other sucking species, may carry dangerous diseases. Others play helpful roles: e.g., pollinating flowers and destroying other insect pests. Their metamorphosis is complete. Many of the worm-like larvae (maggots) pupate in a barrel-shaped puparium.

Housefly *(Musca domestica)*

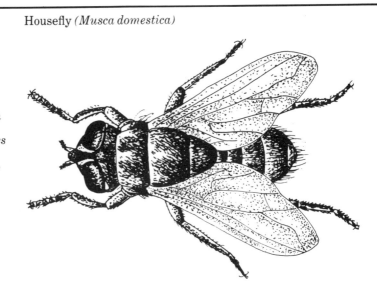

ORDER SIPHONAPTERA—Fleas

Adult fleas are external blood-sucking parasites of birds and mammals. All of the 1,100 known species are wingless and laterally flattened, with long legs used for jumping. They have highly efficient piercing and sucking mouthparts. Their metamorphosis is complete, with the larvae developing as tiny whitish, legless maggots that usually live in or near the host's nest, but as scavengers (not parasites) feeding on organic debris before spinning the cocoons in which they will pupate.

Rodent Flea *(Orchopeas leucopus)*

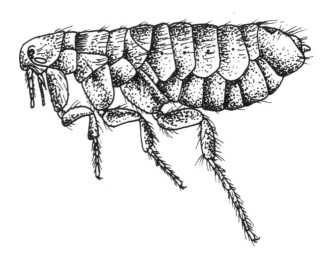

ORDER HYMENOPTERA—Sawflies, Ichneumons, Chalcids, Ants, Wasps, Bees, etc.

This immense and extremely diverse order of insects, with well over 105,000 species and of prime economic importance, includes the most highly developed and complex of all invertebrate species. Those with wings have two membranous pairs: the hindwings are smaller and attached by a series of small hooks to form one contiguous unit in flight. The metamorphosis is complete; in some species, care of the young is remarkably complex in organization.

Velvet Ant *(Dasymutilla occidentalis)*

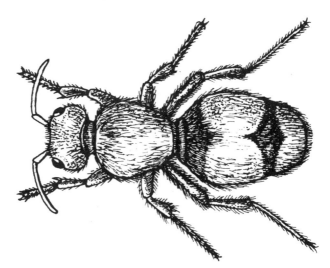

Glossary

Adaptation. An evolutionary modification that fits an organism more perfectly for existence in its environment.

Afforestation. The establishment of forest, by man, on unforested land—usually to yield timber, provide shelter for wildlife, or stop erosion.

Alpine zone. The biogeographic zone of slopes above the timberline, characterized by rosette-forming herbaceous plants and slow-growing woody shrubs. *See:* Foothill zone; Montane zone; Subalpine zone.

Angiosperm. A member of a group of plants (the Angiospermae) in which the seeds are enclosed in an ovary; the great majority of tree species are angiosperms. *See:* Gymnosperm.

Annual. A plant that completes its life cycle during a single growing season.

Annual ring. A circular layer of wood formed each year beneath a tree's bark; its color and thickness indicate the nature of each year's growing season.

Anthocyanins. Pigments that produce most of the blue and red color in plant parts, including leaves and flowers; responsible for most of the beautiful colors of autumn foliage.

Association. A group of plant species that regularly occur together. A forest association is named for its dominant tree species or association; in eastern North America, for example, oak-hickory is a common association, consisting primarily of oaks and hickories.

Bald. A grassy or shrubby area without trees on a ridge or mountain. The origin of balds on certain ridges and shoulders in the southern Appalachians is uncertain, but they may have resulted from early lightning fires or from being cleared for pasture.

Biennial. A plant that lives during two growing seasons, producing only leaves in the first season, flowers and seeds during the second.

Biomass. The amount of living organisms in a given area or habitat, expressed as weight or volume of organisms per unit area (or volume) of the environment.

Boreal forest. A forest of the far north, dominated by conifers.

Broadleaf. Plants having wide-bladed leaves, like those of oaks and maples, rather than needle-like leaves as in conifers.

Canopy. The more or less closed layer formed by the topmost foliage of the tallest trees in a forest.

Carotenoids. Pigments that produce yellow, orange, and certain red hues in plants.

Chaparral. A dense thicket of scrubs or dwarf trees usually found in semiarid climates, especially in California. *See:* Maquis.

Chlorophyll. Pigments that produce the green color of plants; essential to photosynthesis. *See:* Photosynthesis.

Climax. The final stage in which a community of organisms establishes equilibrium with the environment; often the ultimate form of vegetation in an area, given specific environmental conditions of soil and climate. *See:* Sere.

Cloud forest. Dense forest, especially on coastal slopes or in rainy and misty regions at low latitudes; differs from tropical rain forest in being generally denser, cooler, and more humid. *See:* Tropical rain forest.

Clubmoss. A large group of low-growing primitive plants, some species of which produce spores in club-shaped structures called strobiles (also called lycopodium).

Cone. A flower or fruit composed of overlapping, often woody scales; especially, the fruit of pines, spruces, cypresses, and many other needle-leaved trees.

Conifer. A gymnospermous tree that bears its seeds in cones. *See* Evergreen.

Cycad. A primitive gymnospermous plant resembling a palm, in which both male and female flowers are cones; a connective link between the ferns and the conifers.

Deciduous. Characterized by a regular, periodic loss of leaves. Examples of deciduous trees are elms and maples; a few conifers, such as larch and bald cypress, are also deciduous.

Decomposer. A plant that feeds on dead materials and thus causes their mechanical or chemical breakdown; bacteria and fungi are among the agents that decompose forest litter.

Dendrology. The scientific study of trees, as distinguished from other plants.

Detritus. Loose material on the forest floor produced by decay of organic material and by processes of disintegration and erosion.

Ecological succession. The gradual natural replacement of one association of plant and animal species by another, in response to a changing environment. *See:* Sere.

Ecosystem. The entire community of plants and animals in a particular environment and the interrelationships among them.

Endemic. Native to and, under natural conditions, found only in a certain restricted geographic area; for example, the Norfolk Island "pine" is endemic to Norfolk Island, east of Australia, although it has been introduced into other parts of the world by man.

Epiphyte. A plant that grows on another plant; for example, many orchids, bromeliads, and tropical ferns.

Eucalyptus. A genus of evergreen timber trees, rarely shrubs, mostly native to Australia and yielding gums, resins, oils, and tannins as well as wood.

Evergreen. A plant that does not shed all its leaves at one time. Some broad-leaved trees, such as live oak, remain green all year; but most North American and European evergreens are conifers. *See:* Conifer.

Exotic. A plant that is not native to the area where it occurs but is present as a result of being introduced by man.

Foehn. A warm dry wind blowing down a mountainside.

Foothill zone. The lowest band of vegetation in a mountainous region, differing from that of surrounding lowlands as well as from that at higher elevations. *See:* Alpine zone; Montane zone; Subalpine zone.

Forest floor. The richly organic layer of decaying debris lying on the soil.

Frass. The debris or excrement produced by insects when feeding.

Gall. A growth resulting from chemical or mechanical irritation of plant tissues; most galls are caused by insects.

Gallery forest. Forest that grows along a watercourse in a region otherwise largely devoid of trees, as in African and South American savannas.

Gymnosperm. A member of a group of plants (the Gymnospermae) in which the seed is naked, not enclosed in an ovary; group includes the gingko, cycads, and conifers. *See:* Angiosperm.

Hammock (Hummock). A slightly elevated community of dense tropical growth that forms an island in marshy areas of the southern United States; especially, an island of such tropical undergrowth in the Florida Everglades.

Heath. An open area, usually with poor soil and drainage, so called because plants of the Ericaceae (heath) family often predominate.

Herb layer. The lowest layer of plants in a forest, consisting of herbaceous species.

Humus. The topmost layer of soil, consisting of rich decomposing organic material. *See:* Mor; Mull.

Layer. A level of more or less concentrated plant growth at an approximately uniform height above the forest floor. *See:* Canopy; Forest floor; Herb layer; Shrub layer; Understory.

Leader. The main shoot growing from the top of a tree that has a single main trunk.

Liana. A woody vine with long rope-like stems, common in tropical forests.

Mallee (Mallee box). Any of several low-growing Australian eucalypts, or the dense thicket formed by these plants.

Maquis. Dense scrubby underbrush profuse along Mediterranean shores, including many aromatic shrubs. *See:* Chaparral.

Mixed forest. Forest that includes both coniferous and deciduous trees.

Monsoon (From Arabic, meaning "a season.") A wind system that covers a large region and reverses its direction seasonally; first applied to winds over the Arabian Sea that blow for six months from the northeast and for six months from the southwest, producing wet and dry seasons alternately.

Monsoon forest. An open deciduous forest in tropical regions where heavily seasonal rains alternate with prolonged drought.

Montane zone. The band of vegetation that occurs at intermediate elevations in mountainous regions, between foothill and subalpine zones. *See:* Alpine zone; Foothill zone; Subalpine zone.

Moor. A boggy area usually covered with grasses and sedges growing on a layer of peat.

Mor. A type of forest soil formed by a thick, dense mat of slowly decomposing matter, often consisting of conifer needles.

Mulga. Broadly, any Australian tree of scrubby growth and notably hard wood. "Mulga country" is arid land on which mulga is the chief vegetation.

Mull. A type of forest soil in which matter, usually broad leaves, decays rapidly and is mixed thoroughly with animal excretions; it constitutes a rich growing medium.

Muskeg. A mossy bog in the boreal forest region.

Old field. Land formerly cultivated but now untended and grown up with brush.

Old growth. Timber growing in or harvested from a mature forest.

Parasite. A plant or animal that lives in or on another living thing (its host) and obtains part or all of its nutrients from the host's body; for example, mistletoe.

Perennial. A plant that lives for more than two growing seasons, often producing flowers and seeds in several successive years.

Photosynthesis. The process by which green plants synthesize carbohydrates from carbon dioxide and water in the presence of sunlight; the primary process in the production of organic material from inorganic. *See* Chlorophyll.

Rain forest. Forest growing in a zone of rainfall exceeding 2,500 millimeters per year.

Rain shadow. An area located on the leeward side of a mountain barrier and receiving little rainfall.

Rate of growth. The speed at which a tree increases in size, which may be measured radially in the trunk, in lumber cut from the trunk, or in the dimension of the crown or other tree part. A unit of measure in wood is the number of annual rings per 2.54 cm (1 inch).

Relict association. A group of plant species living together and surviving from an earlier geological period, often occurring in widely separated areas; for example, the temperate deciduous forest that formerly occurred across much of the Northern Hemisphere but now survives only in eastern North America and eastern Asia.

Sapling. A young tree.

Saprophyte. A plant, often lacking chlorophyll, that lives on dead organic matter.

Sapwood. The newly formed, light-colored wood that comprises the outer girth of a tree trunk; composed of living xylem tissue that carries water throughout the tree.

Savanna. An open grassland, usually including scattered trees or shrubs, with heavy rainfall alternating with a dry season.

Scrub. Vegetation consisting largely of stunted trees and shrubs, often forming thick stands on poor soil.

Second growth. Timber that has grown up after the clearing of all or much of the previous stand.

Sere. A series of ecological communities that follow one another in a natural succession, as in the change from a meadow to a forest. *See:* Climax.

Shrub layer (Shrub story). A layer of undergrowth, shrubs, and seedling trees ranging from 2.7 to 13.5 m tall; sometimes almost impassable, but sometimes sparse or absent, as in dense fir woods, where the ground, heavily carpeted with fallen needles, may have no undergrowth at all.

Solar radiation. The energy from sunlight that promotes plant growth and sustains living organisms in wooded and meadow ecosystems.

Springwood (Earlywood). Wood formed early in the growing season, usually recognizable as the paler, less dense inner portion of an annual growth ring. *See:* Summerwood.

Stand. A continuous growth or plantation of trees, often of a single species.

Stratification. The occurrence of vegetation in well-defined vertical layers, such as trees, shrubs, and herbs.

Subalpine zone. The band of vegetation that occurs just below the timberline and alpine zone in mountainous regions.

Subtropical rain forest. Rain forest in regions bordering the tropical zone.

Summerwood (Latewood). Wood formed during the summer, usually recognizable as the darker, denser outer portion of an annual growth ring. *See:* Springwood.

Sustained yield. A forestry term for the ideal situation in which timber can be produced in commercially useful quantities year after year.

Symbiosis. (Literally, "living together.") A relationship between two species in which one or both are benefited and neither is harmed.

Taiga. The coniferous forest that stretches across the Northern Hemisphere, especially the moist forests dominated by spruces and firs tundra that extend into warmer regions.

Taproot. A stout anchor root that provides a tree with stability.

Temperate forest. Wooded region of moderate climate.

Temperate rain forest. Woodland found in usually rather mild climatic areas with heavy rainfall. It commonly includes many kinds of trees, but is distinguished from tropical rain forest by the dominant presence of a single type of tree.

Timberline. The upper limit of trees on mountains or the northern limit of forest in high latitudes.

Transpiration. The loss of water from a plant by evaporation, especially through tiny openings in the leaves called stomata.

Tree line. *See:* Timberline.

Tropical rain forest. Forest generally found between the Tropics of Cancer and Capricorn, with much rain and dense growth.

Tundra. Treeless vegetation adapted to long winters and low temperatures. Arctic tundra extends north of the boreal forest; alpine tundra extends above timberline on mountains.

Understory. A layer of plants adapted to growth beneath the main forest canopy.

Uniform stand. A cluster of trees of similar size and age.

Virgin growth. The primeval growth of mature trees.

Wetlands. Moist or flooded lands such as bogs, marshes, and swamps.

Woods. (1) Synonym for forest, except that it sometimes denotes an area of tree growth smaller than a forest. (2) Also, the hardened fibrous material of which the trunks and branches of trees and shrubs are composed.

Index

Page numbers in bold face type indicate photographs

226